计算机类技能型理实一体化新形态系列

Java 面向对象程序设计

（微课视频版）

主 编 吴绍根 吴 边

清华大学出版社
北京

内 容 简 介

本书是一本介绍 Java 面向对象程序设计的基础书籍,知识同步到 Java 最新长期支持版 LTS 17 版,适合 Java 初学者使用。本书共 16 章,全面介绍了 Java 面向对象程序设计的基本概念、基本方法、基本技术和应用实践。第 1 章介绍了 Java 的特点和建立 Java 开发环境;第 2 章和第 3 章介绍了 Java 的基本运算、基本输入/输出和程序流程控制;第 4~7 章介绍了 Java 面向对象程序设计的核心知识,包括类、对象、继承、多态、接口及枚举;第 8~11 章介绍了 JDK 基本类的使用,包括 Java 基础类、异常类、集合类、流式编程、I/O 流,同时,还介绍了 Java 程序的跟踪调试技术;第 12 章介绍了 Java 的反射技术和注解编程;第 13~15 章介绍了 Java 多线程编程、网络编程和数据库编程;第 16 章介绍了 Java 图形界面编程。

本书既可作为高等院校计算机相关专业的教材,也可作为 Java 编程爱好者的自学书籍。

本书封面贴有清华大学出版社防伪标签,无标签者不得销售。
版权所有,侵权必究。举报:010-62782989,beiqinquan@tup.tsinghua.edu.cn。

图书在版编目(CIP)数据

Java 面向对象程序设计:微课视频版 / 吴绍根,吴边主编. -- 北京:清华大学出版社,2024.9(2025.3重印). --(计算机类技能型理实一体化新形态系列). -- ISBN 978-7-302-67178-7

Ⅰ. TP312.8

中国国家版本馆 CIP 数据核字第 2024TS2705 号

责任编辑:张龙卿 李慧恬
封面设计:刘代书 陈昊靓
责任校对:袁 芳
责任印制:宋 林

出版发行:清华大学出版社
网　　址:https://www.tup.com.cn,https://www.wqxuetang.com
地　　址:北京清华大学学研大厦 A 座　　邮　编:100084
社 总 机:010-83470000　　邮　购:010-62786544
投稿与读者服务:010-62776969,c-service@tup.tsinghua.edu.cn
质量反馈:010-62772015,zhiliang@tup.tsinghua.edu.cn
课件下载:https://www.tup.com.cn,010-83470410
印 装 者:三河市龙大印装有限公司
经　　销:全国新华书店
开　　本:185mm×260mm　　印　张:19.5　　字　数:470 千字
版　　次:2024 年 9 月第 1 版　　印　次:2025 年 3 月第 2 次印刷
定　　价:59.80 元

产品编号:102001-01

前　言

　　Java 是一个庞大的技术体系，是目前设计和开发大中小型因特网应用系统的主流技术。Java 自诞生以来，由于其严谨、面向对象、灵活、平台无关等特点，得到了广大使用者的喜爱和广泛应用。

　　从体系结构上看，Java 包括以下几个层次：Java 标准版（Java Standard Edition，Java SE）、Java 企业版（Java Enterprise Edition，Java EE，Java Web 是 Java EE 中的一个主要部分）、Java 框架（Java Spring Framework）、Java 微服务（Java Micro Service）和 Java 云端开发部署（Java Cloud）。其中，基于 Java 可以开发 Java 桌面应用程序。例如，目前流行的 IntelliJ IDEA 开发环境就是基于 Java 技术开发的；结合 Java EE、Java Spring Framework、Java Micro Service 和 Java Cloud 技术，可以开发各种规模的因特网应用系统。例如，目前一些典型的网络应用系统，包括电子政务系统、电子商务系统等都是基于这些技术体系开发和建设的。

　　为了帮助 Java 技术学习者和使用者学习、掌握和使用 Java 技术体系设计和开发应用系统，需要结合现今及未来应用需求，从体系上考虑和合理选取知识内容及安排先后关系，从而在未来实际项目开发中合理应用。清华大学出版社携手院校和企业有经验的教师和工程师开发了一整套 Java 技术体系丛书，本丛书共 5 本，包括《Java 面向对象程序设计（微课视频版）》《Java Web 程序设计（微课视频版）》《Spring 框架应用开发——基于 Spring Boot（微课视频版）》《Spring Cloud 微服务应用开发——基于 Alibaba Nacos（微课视频版）》《Spring 微服务系统部署（微课视频版）》。本书是这套丛书的第 1 本，介绍 Java 核心基础。

　　在 Java 体系的各层次技术架构中，Java 核心基础是支撑后续各种 Java 技术的基石。在对 Java 核心的学习过程中，知识内容的选取、学习和掌握程度会直接影响后续知识的学习、理解和掌握程度。同时，近些年来由于 Java 的广泛使用，其核心技术不断吸纳新的建议而持续向前发展。在编写本书时，Java SE 的最新版本是 JDK 20 版本，其长期支持版也升级到了 LTS 17 版本。因此，融合最新技术发展及后续课程学习需要，本书对 Java 核心基础知识内容做了精心选择，在帮助学习者学习 Java 核心技术的同时，也有助于他们对 Java 体系中后续课程的学习和未来的开发实践中应用 Java 技术。

　　本书共 16 章，全面介绍了 Java 技术的核心基础。学习 Java 知识的目的是在工程中使用 Java 编写规模应用。本书的典型特点是：除了第 1 章、

第 2 章外，其他各章均安排了一节"案例"，通过具体案例介绍各章所述知识点的具体应用；同时，每章还安排了"应用实践"，对各章所述知识点的最佳使用场景和最佳使用方法做了介绍；此外，每章均配有相应的练习题。

第 1 章用简洁的语言介绍了 Java 的特点以及建立 Java 开发环境的方法，本书采用目前较为流行的最新版 IntelliJ IDEA 作为 Java 开发环境；第 2 章介绍了 Java 变量的概念及其使用、Java 基本运算符的使用，以及如何在 Java 程序中进行基于键盘和显示器的数据输入/输出，为后续内容做好准备；第 3 章介绍了 Java 程序的基本流程控制语句，包括分支控制、循环控制、switch 表达式的使用、数组的使用等；第 4 章用通俗易懂的语言化抽象为具体，介绍了 Java 面向对象最为基本的概念，包括如何定义类以及如何创建和使用对象等；第 5 章介绍面向对象的最为核心的概念，包括什么是继承、什么是多态以及如何在工程实践中使用这些技术，同时结合 Java 的最新发展，介绍了 record 及 sealed 的使用；第 6 章介绍了接口的定义和如何实现接口，进一步结合 Java 发展，介绍了函数式接口和 lambda 表达式的使用；第 7 章介绍了如何定义和使用枚举类型，并给出了使用枚举类型的实践场景；第 8 章选取了 Java JDK 中常用的类进行介绍，包括基本类、String 类等，同时介绍了经常被忽略但是很重要的字符编码的概念；第 9 章介绍的异常处理是 Java 程序中不可或缺的部分，同时，Java 程序的调试和跟踪技术在工程实践中会被经常用到，因此，本章专门开辟了一节介绍如何调试和跟踪 Java 程序；第 10 章对集合类和流式编程做了详细介绍；第 11 章对 Java 的 I/O 进行仔细梳理，对在工程中使用非常广泛的 10 个类进行介绍，从而化复杂为简单；第 12 章通过案例简洁地介绍了 Java 反射和注解编程；第 13 章介绍了多线程的概念、如何创建 Java 线程、线程并发控制技术、生产者—消费者模型及线程池的使用；第 14 章介绍了如何使用 Java 进行基于 TCP/IP 的网络通信；第 15 章介绍了如何在 Java 程序中使用标准的 JDBC 驱动访问和操作数据库数据；第 16 章介绍了如何基于 Java Swing 编写图形界面程序。

将本书作为高校计算机相关专业的教材使用时，建议授课课时安排在 72 课时左右，当然，各个院校也可根据各自情况进行适当的调整。

本书的第 1 章、第 2 章由吴边编写，第 3~16 章由吴绍根编写。本书配有详细的 PPT、源代码、课后练习解答等资源，这些资源可从清华大学出版社官网下载。教学视频可扫描二维码学习。

由于编者水平有限，书中难免存在疏漏之处，敬请读者提出宝贵意见。

编　者

2024 年 4 月

目　　录

第 1 章　建立 Java 程序开发环境 ·· 1
1.1　Java 语言概述 ·· 1
1.1.1　程序设计语言 ·· 1
1.1.2　Java 语言的特点 ··· 1
1.2　建立 Java 开发环境 ··· 2
1.3　第一个"Hello world!"程序 ··· 2
1.3.1　创建 Java 程序工程 ·· 2
1.3.2　运行 Java 程序 ·· 5
1.4　Java 程序的运行过程 ·· 6
1.4.1　编译代码 ·· 7
1.4.2　Java 程序的运行机理 ·· 8
1.5　练习：安装 Java 开发环境 ·· 9

第 2 章　Java 基本运算和输入 / 输出 ····································· 10
2.1　Java 程序的组成 ·· 10
2.2　Java 基本数据类型和字面常量 ··· 11
2.3　变量 ·· 11
2.3.1　定义变量和访问变量 ··· 12
2.3.2　显示变量的值 ·· 12
2.4　数据运算 ·· 13
2.5　Java 基本输入 / 输出和 String 类的使用 ····························· 13
2.5.1　基本输出语句 ·· 14
2.5.2　基本输入语句 ·· 16
2.5.3　String 类的使用 ··· 17
2.6　练习：计算工资 ·· 18

第 3 章　Java 程序流程控制 ·· 19
3.1　顺序语句和 if 分支语句 ·· 19
3.2　switch...case default 分支语句及其应用实践 ······················· 19
3.3　循环语句 ·· 21
3.4　数组 ·· 21
3.4.1　定义数组 ·· 22

3.4.2 访问数组元素 ··· 22
3.4.3 使用 for each 遍历数组元素 ··· 23
3.4.4 二维数组 ··· 24
3.5 switch 表达式和 yield 关键字的使用 ··· 26
3.6 函数及其调用 ··· 27
3.7 案例：学生成绩计算系统 ··· 29
3.7.1 案例任务 ··· 29
3.7.2 任务分析 ··· 29
3.7.3 任务实施 ··· 30
3.8 练习：计算质数及其和 ··· 30

第 4 章 类和对象 ··· 31
4.1 定义类和创建对象 ··· 31
4.1.1 类的含义 ··· 31
4.1.2 定义类 ··· 32
4.1.3 在 IDEA 中创建 Java 类 ··· 34
4.1.4 创建及使用对象 ··· 37
4.2 构造函数 ··· 41
4.2.1 类的构造函数 ··· 41
4.2.2 构造函数重载 ··· 43
4.3 类的静态属性、静态方法和静态代码块 ··· 45
4.3.1 静态属性 ··· 46
4.3.2 静态方法 ··· 48
4.3.3 静态代码块 ··· 49
4.3.4 静态属性、静态方法和静态代码块应用实践 ································· 51
4.4 内部类 ··· 51
4.4.1 成员内部类 ··· 51
4.4.2 静态内部类 ··· 54
4.4.3 使用内部类应用实践 ··· 56
4.5 案例：使用 Java 类描述一元二次方程 ··· 56
4.5.1 案例任务 ··· 56
4.5.2 任务分析 ··· 56
4.5.3 任务实施 ··· 56
4.6 练习：计算三角形的面积和周长 ··· 58

第 5 章 继承和多态 ··· 59
5.1 类的继承 ··· 59
5.1.1 继承的概念 ··· 59
5.1.2 定义类的继承关系 ··· 60

目 录

  5.1.3 super 关键字及方法重写 ············ 64
  5.1.4 练习：完成 Teacher 子类和 Worker 子类的代码编写 ············ 66
 5.2 访问限定符 ············ 66
  5.2.1 访问限定符及其可访问性 ············ 66
  5.2.2 访问限定符使用举例 ············ 67
 5.3 抽象类和多态 ············ 67
  5.3.1 抽象类和使用 final 关键字修饰属性 ············ 67
  5.3.2 多态 ············ 71
  5.3.3 使用 instanceof 关键字检查对象类型 ············ 72
  5.3.4 对象数组 ············ 73
 5.4 使用 final、record 和 sealed 关键字修饰类 ············ 76
  5.4.1 使用 final 关键字修饰类 ············ 76
  5.4.2 使用 record 关键字定义 Java 类 ············ 76
  5.4.3 使用 sealed 关键字修饰类 ············ 78
 5.5 案例：定义 Java 程序类应用实践 ············ 78
  5.5.1 案例任务 ············ 78
  5.5.2 任务分析 ············ 78
  5.5.3 任务实施 ············ 79
 5.6 练习：打印自定义图形形状 ············ 82

第 6 章 接口 ············ 83

 6.1 接口及其应用 ············ 83
  6.1.1 定义接口 ············ 84
  6.1.2 实现接口 ············ 85
  6.1.3 使用接口及 instanceof 关键字在接口中的应用 ············ 87
  6.1.4 接口的继承 ············ 89
 6.2 接口的默认方法、静态方法和私有方法 ············ 90
 6.3 函数式接口和 lambda 表达式 ············ 93
  6.3.1 函数式接口 ············ 93
  6.3.2 使用匿名内部类实现接口 ············ 94
  6.3.3 lambda 入门：使用 lambda 表达式实现函数式接口 ············ 95
  6.3.4 lambda 表达式基本语法 ············ 96
  6.3.5 接口方法引用 ············ 97
 6.4 接口、匿名内部类和 lambda 表达式应用实践 ············ 97
 6.5 案例：按价格排序不同产品 ············ 97
  6.5.1 案例任务 ············ 98
  6.5.2 任务分析 ············ 98
  6.5.3 任务实施 ············ 98
 6.6 练习：计算空间中两点的距离 ············ 103

第 7 章 枚举类型 ... 104
7.1 枚举类型入门：一个表示四季的枚举类型 ... 104
7.2 枚举类型进阶 ... 106
7.3 枚举类型应用实践 ... 108
7.4 案例：员工 Staff 类 ... 109
7.4.1 案例任务 ... 110
7.4.2 任务分析 ... 110
7.4.3 任务实施 ... 110
7.5 练习：水果的成熟季节 ... 111

第 8 章 Java 基础类的使用 ... 112
8.1 Java 基本类 ... 112
8.1.1 Object 类 ... 112
8.1.2 基本数据类型的包装类 ... 113
8.1.3 大数据类 ... 114
8.1.4 System 类 ... 115
8.1.5 Math 类 ... 116
8.2 字符串类 ... 116
8.2.1 String 类 ... 117
8.2.2 StringBuffer 类 ... 117
8.3 随机数生成器类 ... 119
8.3.1 Random 类 ... 119
8.3.2 使用 RandomGenerator 接口生成随机数 ... 120
8.4 日期时间类 ... 121
8.4.1 Date 类 ... 121
8.4.2 Calendar 类 ... 121
8.4.3 SimpleDateFormat 类 ... 122
8.5 使用 Java 基础类应用实践 ... 124
8.6 案例：猜数游戏 ... 124
8.6.1 案例任务 ... 124
8.6.2 任务分析 ... 124
8.6.3 任务实施 ... 124
8.7 练习：计算闰年 ... 126

第 9 章 Java 程序异常及程序调试技术 ... 127
9.1 程序错误分类 ... 127
9.2 Java 程序异常及其处理入门 ... 127
9.2.1 Java 程序异常现象举例 ... 127
9.2.2 Java 异常处理入门 ... 129

9.3　Java 程序异常及其处理进阶 131
　　9.3.1　Error 类 132
　　9.3.2　Exception 类 133
　　9.3.3　非检查性异常 133
　　9.3.4　检查性异常 133
　　9.3.5　Java 异常处理 135
　　9.3.6　自定义异常 135
9.4　案例：处理程序异常 136
　　9.4.1　案例任务 136
　　9.4.2　任务分析 136
　　9.4.3　任务实施 137
9.5　在 IDEA 中调试 Java 程序 140
9.6　Java 异常及程序调试应用实践 142
9.7　练习：将从键盘输入的字符串转换为浮点数 142

第 10 章　集合类及流式编程 143

10.1　泛型 143
　　10.1.1　泛型入门 143
　　10.1.2　泛型类 145
　　10.1.3　泛型方法 146
　　10.1.4　泛型接口 147
　　10.1.5　泛型类型限制和泛型通配符"？" 150
10.2　集合类 150
　　10.2.1　集合类主要接口和类之间的关系 150
　　10.2.2　List 接口及其实现类的使用 151
　　10.2.3　Set 接口及其实现类的使用 154
　　10.2.4　Map 接口及其实现类的使用 158
　　10.2.5　数组工具类 Arrays 的使用 160
10.3　Java 流式编程 161
　　10.3.1　Java 常用函数式接口及其使用 161
　　10.3.2　Optional 类及泛型通配符"？"使用举例 164
　　10.3.3　流式编程入门 168
　　10.3.4　创建 Stream 和操作 Stream 170
10.4　Java 数组、集合类及流式编程应用实践 170
10.5　案例：自制词典 171
　　10.5.1　案例任务 171
　　10.5.2　任务分析 171
　　10.5.3　任务实施 171
10.6　练习：使用流式编程查询学生信息 172

第 11 章 文件输入 / 输出操作 173

11.1 文件基本操作 173
11.1.1 使用 File 类操作文件属性 173
11.1.2 使用 Files 类操作文件属性及读 / 写文件内容 175
11.1.3 使用 WatchService 监视目录和文件变化 175

11.2 字节流读 / 写 175
11.2.1 字节流的含义 175
11.2.2 读 / 写文件字节流 176
11.2.3 使用 try-with-resource 处理异常和关闭资源 178
11.2.4 读 / 写内存字节流 180

11.3 字符流读 / 写 181
11.3.1 字符编码和字符解码 182
11.3.2 无缓冲字符流读 / 写 184
11.3.3 带缓冲字符流读 / 写 186

11.4 对象数据读 / 写 188
11.5 Java 流操作应用实践 190
11.6 案例：通讯录程序 190
11.6.1 案例任务 190
11.6.2 任务分析 191
11.6.3 任务实施 191

11.7 练习：自制工资管理程序 195

第 12 章 Java 反射和注解 196

12.1 Java 反射 196
12.1.1 反射概念的引入 196
12.1.2 反射的核心——Class 类 197
12.1.3 通过反射获取类的构造方法、属性和普通方法 199

12.2 Java 注解 202
12.2.1 Java 标准注解 202
12.2.2 自定义注解 202

12.3 Java 反射与注解应用实践 206
12.4 案例：自动注入对象 207
12.4.1 案例任务 207
12.4.2 任务分析 207
12.4.3 任务实施 207

12.5 练习：自动注入 Teacher 对象 210

第 13 章 多线程 211

13.1 Java 多线程入门 211

目　录

- 13.2 Thread 类及创建子线程 214
  - 13.2.1 通过继承 Thread 类创建线程 215
  - 13.2.2 通过实现 Runnable 接口创建线程 215
  - 13.2.3 使用 FutureTask 创建线程 218
- 13.3 线程状态、线程调度和线程优先级 221
- 13.4 线程并发控制 222
  - 13.4.1 多线程中数据的不一致性现象举例 222
  - 13.4.2 使用 synchronized 控制线程并发 225
  - 13.4.3 使用原子类型变量控制线程并发 227
  - 13.4.4 使用 Lock 接口控制线程并发 229
- 13.5 线程同步控制及生产者—消费者模型 229
- 13.6 线程池 229
  - 13.6.1 Java 线程池框架 230
  - 13.6.2 线程池使用举例 231
  - 13.6.3 多例多线程和单例多线程及 ThreadLocal 类的使用 234
- 13.7 Java 线程应用实践 237
- 13.8 案例：找出小于 1000 的所有质数和水仙花数 238
  - 13.8.1 案例任务 238
  - 13.8.2 任务分析 238
  - 13.8.3 任务实施 238
- 13.9 练习：统计上网人数 241

第 14 章　网络编程 **242**

- 14.1 网络通信协议 242
  - 14.1.1 IP 地址 InetAddress 类和端口 242
  - 14.1.2 UDP 和 TCP 244
- 14.2 使用 UDP 进行通信 244
  - 14.2.1 DatagramSocket 类和 DatagramPacket 类 244
  - 14.2.2 UDP 点对点通信程序举例 245
- 14.3 使用 TCP 进行通信 249
  - 14.3.1 客户/服务器模式 249
  - 14.3.2 ServerSocket 类和 Socket 类 249
  - 14.3.3 TCP 通信程序举例 250
- 14.4 使用 HTTP 访问网络页面 254
  - 14.4.1 Java 对 HTTP 的实现概述 254
  - 14.4.2 使用 HttpClient 访问网络页面 256
- 14.5 Java 网络编程应用实践 257
- 14.6 案例：聊天程序 258
  - 14.6.1 案例任务 258

	14.6.2 任务分析	258
	14.6.3 任务实施	258
14.7	练习：完善聊天程序 Chatter 类的代码	263

第 15 章　使用 JDBC 访问数据库　264

- 15.1 JDBC 概述　264
- 15.2 加载数据库驱动程序　265
- 15.3 JDBC 接口访问数据库的核心类和核心接口　266
  - 15.3.1 DriverManager 类注册数据库驱动程序　266
  - 15.3.2 Connection 接口建立与数据库的连接　267
  - 15.3.3 Statement 接口执行 SQL 语句　268
  - 15.3.4 PreparedStatement 接口执行参数化 SQL 语句　269
  - 15.3.5 ResultSet 接口处理查询结果　270
- 15.4 案例：实现对 book 表的增删改查　271
  - 15.4.1 案例任务　271
  - 15.4.2 任务分析　271
  - 15.4.3 任务实施　271
- 15.5 JDBC 应用实践　278
- 15.6 练习：完成案例程序的删改查功能　278

第 16 章　Java 图形用户界面　279

- 16.1 Swing 概述　279
- 16.2 容器组件　279
  - 16.2.1 JFrame 顶级窗口容器　279
  - 16.2.2 JPanel 面板容器　281
- 16.3 布局管理器　283
  - 16.3.1 FlowLayout 布局　283
  - 16.3.2 GridLayout 布局　285
  - 16.3.3 BorderLayout 布局　286
- 16.4 Swing 常用组件　288
- 16.5 Swing 事件处理　290
- 16.6 Java 图形界面应用程序开发实践　293
- 16.7 案例：图像混合器　293
  - 16.7.1 案例任务　293
  - 16.7.2 任务分析　293
  - 16.7.3 任务实施　294
- 16.8 练习：图形界面聊天程序　297

参考文献　298

第 1 章　建立 Java 程序开发环境

Java 是一门重要的程序设计语言。使用 Java 语言既可以开发桌面应用程序，也可以开发基于网络的应用程序。一些知名的电子商务系统、电子政务系统、自动化办公系统等，很多是使用 Java 语言开发的。本章首先简单介绍 Java 语言的特点，然后介绍如何安装 Java 语言程序开发环境，最后使用 Java 语言开发一个简单的程序。

1.1　Java 语言概述

计算机是 20 世纪 40 年代人类最伟大的发明创造之一。计算机从其诞生之日起，为人类的文明进步和发展做出了巨大贡献。近年来，以高性能计算机为基础的大数据和人工智能的发展和应用，使计算机相关技术再次成为技术发展的重要方向。但是，要使计算机按照人们的意愿工作，人们必须使用计算机能够理解的称为"计算机程序设计语言"的工具来完成。

1.1.1　程序设计语言

计算机程序设计语言，简称程序设计语言，是人们对计算机"发号施令"进而控制计算机工作的重要工具。程序设计语言包括高级语言和低级语言：C/C++、Java、Python、C# 等，都是常用的高级语言；机器语言、汇编语言，则是低级语言。这里的"高级语言""低级语言"是针对程序设计语言的语义丰富方面而言的，并不具有日常概念上的层次高低的含义。简单来说，高级语言语义丰富，容易学习和使用，是开发应用程序的主要语言；而低级语言则语义单一，不易学习也不易使用，低级语言是计算机出现初期被使用的主要语言。

1.1.2　Java 语言的特点

Java 语言是高级程序设计语言，是当今非常流行和广为使用的程序设计语言之一。Java 语言功能丰富、容易学习也容易使用。概括起来，Java 语言具有以下特点。

（1）简单：相对于其他高级程序设计语言，Java 语言出现的时间比较晚，因此，Java 从语言的设计上吸纳了其他高级语言的优良性质，并克服了其他语言的缺点和不足，所以，Java 语言简单且易学易用。

（2）面向对象：Java 语言是一种面向对象的编程语言。"面向对象"是当代程序设计语言的一大进步和创举。通过面向对象编程技术，Java 语言将程序的各个部分看作一个个的"对象",通过对象之间的交互达成程序的功能、性能目标。

(3)解释执行或实时执行:Java编译程序编译Java源代码并生成字节码,所生成的字节码既可以通过Java虚拟机直接解释执行,也可以通过Java的JIT(just in time)技术将字节码转换成计算机机器码,从而提高执行效率。

(4)可移植:由于Java编译器将Java源代码编译生成Java字节码,并通过Java虚拟机解释执行,所以,Java程序并不依赖具体执行平台,用Java编写的程序可以在任何操作系统上运行。

(5)支持多线程:Java语言从设计上就支持多线程,并提供了优秀的多线程并发控制机制,因此,相比于其他高级语言,使用Java编写多线程程序更容易也更可靠。

1.2 建立 Java 开发环境

有句老话:工欲善其事必先利其器。要学习Java语言并使用Java语言开发计算机应用程序,首先需要建立Java开发环境。为了建立Java开发环境,需要执行以下两个基本步骤:①安装JDK;②安装IntelliJ IDEA工具。本书使用在业界广泛使用的IntelliJ IDEA作为编程工具环境。

安装JDK和IntelliJ IDEA

安装JDK和IntelliJ IDEA 说明

1.3 第一个"Hello world!"程序

现在,已经安装了开发Java程序所必需的工具,可以开发Java程序了。在开发Java程序之前,需要在IDEA中创建一个Java程序工程。为什么需要创建"程序工程"呢?因为一个Java程序会包括多个文件,同时,还会包括一些以管理为目的的辅助性文件,IDEA工具需要对这些文件需要进行统一集中管理。使用"程序工程"这项技术就可以达到这个管理目标。

1.3.1 创建Java程序工程

下面创建一个最简单的Java程序工程,运行这个程序后,将在显示器上显示"Hello world!",因此,这个程序称为"Hello world!"程序。启动IDEA,单击界面上的"+"按钮,创建一个Java程序工程,如图1-1所示。

在如图1-1所示的界面中,在Name输入框中输入工程的名称;在Location选择框中选择存放工程的文件夹;在Language选项中选择Java;在Build system选项中选择Maven;在JDK选项中选择所安装的JDK工具,

创建Java程序工程

由于之前安装的JDK工具是JDK 17,所以这里选择17 Oracle OpenJDK version 17.0.5;选中Add sample code单选框;单击Create按钮创建这个名称为ch01-01的工程。成功创建

Java 程序工程后，将显示如图 1-2 所示的界面。

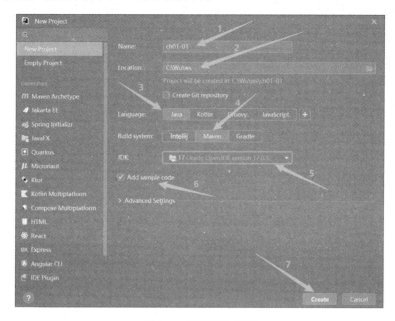

图 1-1　创建程序工程

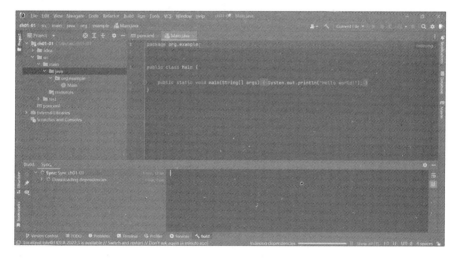

图 1-2　成功创建 Java 程序工程后的界面

由于这是第一次创建 Java 程序工程，IDEA 需要下载必要的软件支持插件，这需要一定的时间，请耐心等待这个过程完成。当图 1-2 中最下方的进度条消失后，表示下载完成，将显示如图 1-3 所示的界面。

在如图 1-2 和图 1-3 所示的界面中，界面的背景颜色太黑，同时，文字字号太小，可以根据需要修改界面的主题和文字大小。为此，在如图 1-3 所示的界面中，选择 File → Settings 命令，如图 1-4 所示。

在如图 1-5 所示的界面中，在 Theme 选择框中选择 IntelliJ Light，通过修改 Size 值设置文字大小。例如，在这里设置为 14。

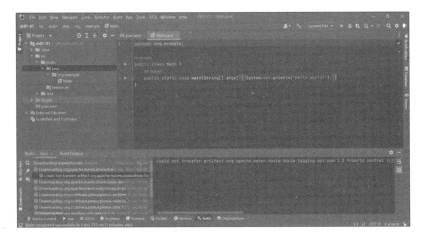

图 1-3　IDEA 完成支持工具下载

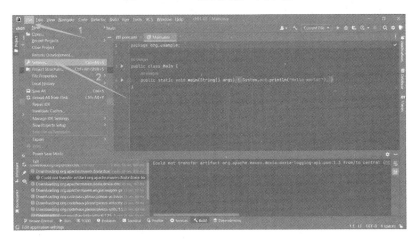

图 1-4　修改 IntelliJ IDEA 界面主题和文字大小

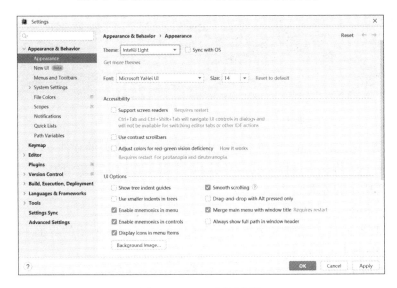

图 1-5　IDEA 界面主题

在如图 1-5 所示的界面中，选择左边的 Editor → Font 命令，如图 1-6 所示。

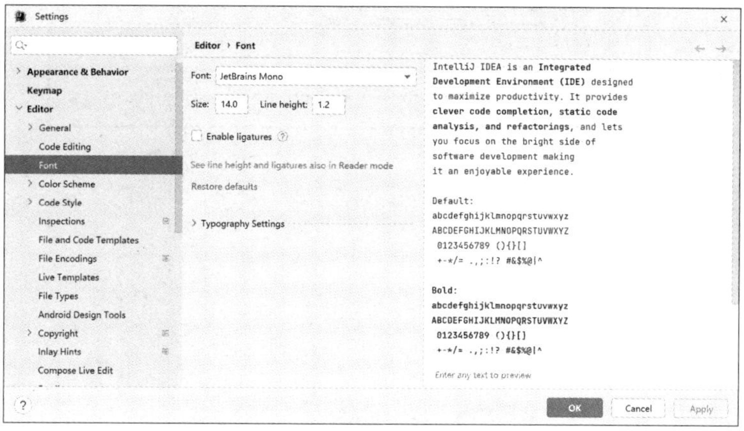

图 1-6　代码编辑文字大小设置

在图 1-6 中，通过修改 Size 值设置文字大小。例如，在这里设置为 14。完成设置后，单击 OK 按钮，返回主界面，如图 1-7 所示。

图 1-7　IDEA 主界面

如图 1-7 所示是程序开发人员使用 IDEA 开发应用程序时的工作界面：在这个界面中，既可以编写 Java 程序代码，也可以运行 Java 程序，还可以调试 Java 程序等。下面先运行这个刚刚创建的 Java 程序看看效果。

1.3.2　运行 Java 程序

由于在如图 1-1 所示的界面中选中了 Add sample code 单选框，所以 IDEA 为这个程序工程创建了一个必需的称为"Java 主类"的代码（包含 main( ) 入口函数的类称为主类）。这个主类代码如图 1-8 所示。矩形框中的代码就是 Java 主类代码，保存在名称为 Main.java 的文件中。矩形框中的代码和保存在 Main.java 中的内容是一致的，它们都被称为"Java 源代码"。任何一个可运行的 Java 程序都必须至少有一个主类代码。

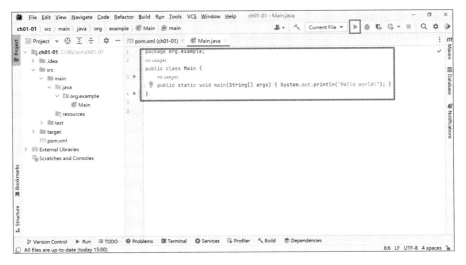

图 1-8　Java 程序主类代码

由于 ch01-01 程序工程已经有了 Java 主类代码，因此，这个程序是可以直接运行的。为此，在如图 1-8 所示的界面中单击▶按钮，IDEA 经过必要的处理后，将显示如图 1-9 所示的运行结果。

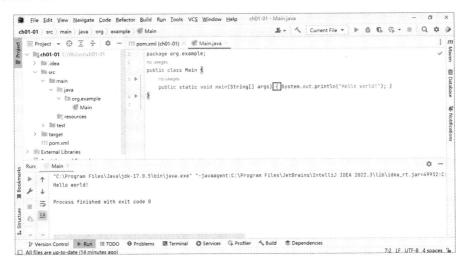

图 1-9　ch01-01 程序工程的运行结果

在如图 1-9 所示的界面中，"Hello world!"就是 ch01-01 程序工程运行后显示的一段文字信息。那么，这个程序为什么会显示这句话呢？这需要了解 Java 程序的运行过程。

1.4　Java 程序的运行过程

在 IDEA 中运行一个 Java 程序非常简单：只需要在如图 1-8 所示的界面中单击▶按钮即可启动 Java 程序运行。其实，在 IDEA 内部，为了运行一个 Java 程序，会经过以下两个阶段：第一个阶段称为编译或者构建阶段，在这个阶段，IDEA 会使用 JDK 中的 Java 编译器将 Java 源代码编译（也称为"构建"，或 Build）成 Java 的字节码；第二个阶段称为执行阶段，IDEA 会使用 JDK 中的 Java 虚拟机执行在第一个阶段所生成的 Java 字节码，进而启动 Java 程序运行。

1.4.1 编译代码

如果源代码没有错误，编译成功的 Java 代码即可被 Java 虚拟机启动运行；如果 Java 源代码有错误，如将某个对象的名字输错了，那么，Java 编译器是无法正确编译 Java 源代码的。例如，在如图 1-10 所示的 Main.java 的源代码中，如果将其中的 out 对象的名字错误地输入成 outt，会发现该程序不能运行，将显示如图 1-10 所示的错误信息。

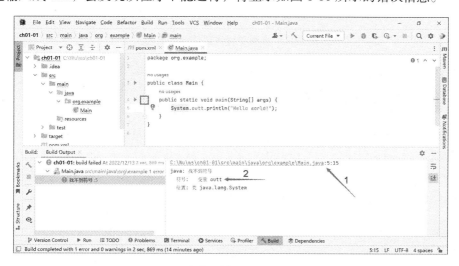

图 1-10　程序代码错误导致不能正确运行

在图 1-10 中，IDEA 将显示如箭头 1 所指示的源代码中的错误，并在箭头 2 所指示的地方给出错误的原因。

细心的读者会发现，如图 1-10 所示的代码与如图 1-9 所示的代码有细微不同：在如图 1-10 所示的代码中，main() 函数的两个大括号和程序语句在不同的行中；而在如图 1-9 所示的代码中，main() 函数的两个大括号和程序语句在同一行。本质上，如图 1-9 所示的代码和如图 1-10 所示的代码是一致的。在如图 1-10 所示的代码中，单击矩形框中的 按钮后，代码的显示效果将与如图 1-9 所示的代码一致；或者，在如图 1-9 所示的代码中，单击矩形框中的大括号"{"后，代码的显示效果将与如图 1-10 所示的代码一致。

另外，如果希望关闭如图 1-10 所示的编译错误信息，可以单击 Build 选项卡关闭它，再次单击 Build 选项卡又可以将编译错误信息显示出来。

还有一种能够直观发现代码错误的方法：将光标放在代码中出现红色文字的地方，即可看到代码错误的原因，如图 1-11 所示。

在图 1-11 中，矩形框内的数字表示代码中存在错误的个数；IDEA 通过小框提示错误信息，例如，此处的错误表示 IDEA 不能解析 outt 这个符号。

图 1-11　直观发现代码错误的方法

1.4.2　Java 程序的运行机理

当 IDEA 使用 Java 编译器成功将 Java 源代码编译为 Java 字节码后，IDEA 会使用 Java 虚拟机执行 Java 程序。为了执行 Java 程序，Java 虚拟机会在 Java 源代码中寻找名称为 main 的入口函数。例如，在如图 1-10 所示的 ch01-01 的 Java 工程中，包含 main( ) 入口函数的代码如下所示（源代码为 01-01.java）。

```
package org.example;
public class Main {
    public static void main(String[] args) {
        System.out.println("Hello world!");
    }
}
```

Java 虚拟机找到 main( ) 入口函数后，会逐条语句地执行由一对大括号所括住的语句。在这个例子中，Java 虚拟机会执行语句：

```
System.out.println("Hello world!");
```

这条语句的作用就是在显示器上显示"Hello world!"文字。这就是为什么会在图 1-9 中显示"Hello world!"。现在，对 Main 主类中的 main( ) 函数做以下修改（源代码为 01-02.java）。

```
package org.example;
public class Main {
    public static void main(String[] args) {
```

```
        System.out.println("Hello world!");
        System.out.println(" 我在学习 Java 面向对象程序设计！");
    }
}
```

当 Java 虚拟机执行这个程序时，会逐条执行 main( ) 函数的大括号所括住的语句，也就是先执行语句：

```
System.out.println("Hello world!");
```

在显示器上显示"Hello world!"文字，然后执行语句：

```
System.out.println(" 我在学习 Java 面向对象程序设计！");
```

在显示器上显示"我在学习 Java 面向对象程序设计！"文字。修改后程序的执行结果如图 1-12 所示。

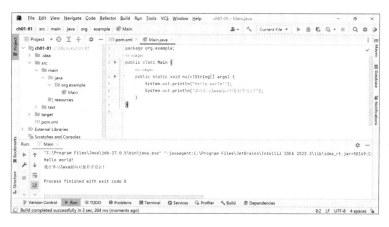

图 1-12　修改后程序的执行结果

1.5　练习：安装 Java 开发环境

请在自己的计算机上安装 Java 程序开发环境，完成后编写一个能够显示如图 1-13 所示四边形的 Java 程序。

图 1-13　四边形

第 2 章　Java 基本运算和输入 / 输出

近年来，计算机在电子商务、电子政务、人工智能和大数据方面得到了广泛应用。但是本质上，计算机就是做数学运算的机器。因此，作为控制计算机工作的 Java 程序，必须具备基本的运算功能，这些功能包括：加减乘除数学运算、大小关系运算、是非（真假）逻辑运算等。本章首先介绍 Java 程序的组成，然后介绍 Java 程序的基本数据类型和如何使用变量保存程序运算数据，最后介绍如何在 Java 程序中进行数据的输入 / 输出。

2.1　Java 程序的组成

Java 程序工程由一个或多个程序包构成，在每个 Java 程序包下，可以存放 0 个或多个 Java 程序类文件（或简称为类）。每个 Java 程序类又是由一个或多个成员属性及函数组成的，每个函数包含 0 条或多条 Java 语句。为了清晰地看到 Java 程序的基本组成结构，下面新建一个名称为 ch02-01 的 Java 程序工程。

为了在已经打开项目工程的 IDEA 中新建一个程序工程，选择 File → New → Project 命令，如图 2-1 所示。

图 2-1　新建一个程序工程

出现的界面与第 1.3 节介绍的创建程序工程的过程一致，只是要注意这里将程序工程的 Name 修改为 ch02-01。创建完成的工程如图 2-2 所示。

图 2-2　创建完成的工程

单击 Main 选项，即可显示 Main.java 文件的内容，它是 Main 类的 Java 源代码，如图 2-3 所示。

图 2-3　Java 程序基本结构

每个程序包都必须有唯一的名字，例如，如图 2-3 所示的程序包的名称为 org.example。一般而言，一个 Java 程序工程会包含多个程序包，每个程序包既可以包含多个 Java 源代码，也可以包含多个子程序包。程序包为管理包含多个 Java 源代码的程序工程带来了便利。程序包的名字一般形如 ×××.×××.××× 的形式，其中的 ××× 为任意合法的英文单词、单词缩写或者中文拼音等。关于程序包的一个编程应用实践是：合理规划程序包的命名，并将你编写的每个 Java 源代码放置在你认为合适的程序包中。

在图 2-3 中，Main 称为 Java 程序类，简称为类。在这里，这个 Java 类的名称为 Main。Java 是面向程序设计语言，Java 类是创建 Java 对象的基础。有关类和对象的概念将在后续章节详细介绍。图 2-3 中的 Main 类也是 ch02-01 程序工程的主类。

在 Main 主类中，包含一个 main 主函数，注意，这里的 main 中的 m 是小写的，而 Main 类中的 M 是大写的。在 Java 中，大写字母和小写字母是不同的。将大写和小写字母视为不同的符号称为大小写敏感，也称为 case sensitive。在 main( ) 函数中，可以包含一条或多条语句。在 Java 程序中，每条语句都必须以英文分号"；"结束。注意，这个 main( ) 函数非常重要，它是 Java 程序的入口：当 Java 虚拟机执行 Java 程序时，就是从 main( ) 函数的第一条语句开始逐条执行包含在其中的语句的。

2.2　Java 基本数据类型和字面常量

Java 提供了完备的基本数据类型和字面常量。Java 的基本数据类型包括 byte、short、int、long、float、double、char、boolean 等；Java 的字面常量包括：整数常量、浮点常量、布尔常量、字符常量和字符串常量等。

Java 基本数据类型和字面常量

2.3　变　　量

为了在程序中保存被处理的原始数据和处理后的结果数据，需要使用变量。变量在任何程序设计语言中都是一个非常重要的概念。

2.3.1 定义变量和访问变量

定义变量和访问变量

变量在可以被使用前必须被定义（或称为声明）。在定义变量时需要明确指明变量的名字和保存在变量中的数据类型。

2.3.2 显示变量的值

程序设计的核心任务是对原始数据进行处理，并将处理的结果显示出来。可以将要处理的原始数据保存在变量中，在继续介绍如何对数据进行处理前，先介绍一下如何将保存在变量中的数据显示出来。目前，ch02-01 程序工程中 Main 主类的 main( ) 入口函数代码如图 2-4 所示。

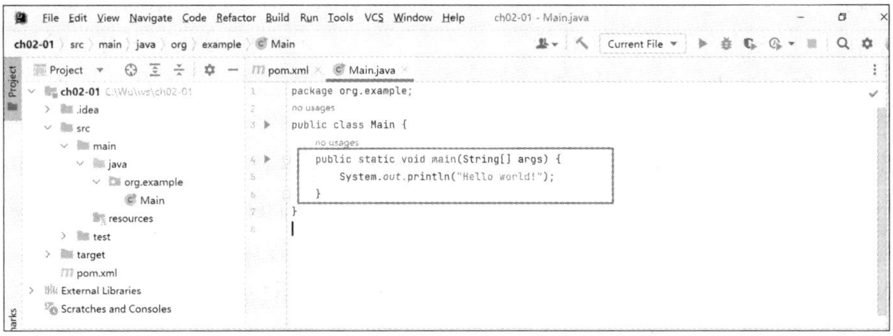

图 2-4 ch02-01 程序工程中 Main 主类的 main( ) 入口函数代码

图 2-4 中，矩形框中的代码是新建工程时的 main( ) 入口函数的代码，现在，修改其中的代码为以下代码（源代码为 02-01.java）。

```java
package org.example;
public class Main {
    public static void main(String[] args) {
        int count;
        count = 100;
        double salary = 5218.5;
        System.out.println(count);
        System.out.println(salary);
    }
}
```

其中的两条语句：

```java
System.out.println(count);
System.out.println(salary);
```

就是在显示器上显示变量 count 和变量 salary 的值。现在运行这个程序,结果如图 2-5 所示。

图 2-5　显示变量的值

在图 2-5 中,用矩形框框住的是修改后的代码,圆圈内的数字显示了存储在变量值中的数据值。

2.4　数 据 运 算

数据运算

计算机的本职工作就是做数据运算。这些运算包括:加减乘除算术运算、关系比较运算、真假逻辑运算等。

2.5　Java 基本输入 / 输出和 String 类的使用

为了在显示器上显示一个信息,例如,为了显示一个常量,或者显示一个变量的值,采用了以下语句。

```
int count = 100;
float salary = 6500.5F;
System.out.println("Hello, World!");
System.out.println(count);
System.out.println(salary);
```

本节对 Java 的基本输入 / 输出语句进行介绍,同时对 Java 中用于表示字符串数据的 String 类的使用进行简单介绍。

2.5.1 基本输出语句

System.out.println(...) 是 Java 中基本的输出语句。这条语句将小括号中的值(称为参数)显示在屏幕上,并在显示参数值后自动换行。与其相对应的输出语句 System.out.print(...) 也用于显示小括号中的参数信息,但是显示参数信息后不换行。例如,修改 Main 类的代码为以下代码(源代码为 02-02.java)。

```java
package org.example;
public class Main {
    public static void main(String[] args) {
        int count = 100;
        long distance = 999910000000L;
        boolean logic;
        logic = true;
        System.out.println(count);
        System.out.print(distance);
        System.out.print(logic);
    }
}
```

运行这个程序的结果如图 2-6 所示。

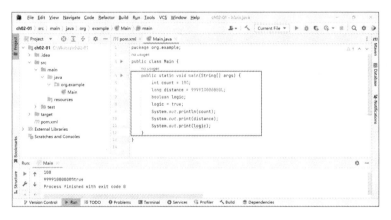

图 2-6 运行基本输出语句的结果

在图 2-6 中,发现输出的 distance 的值与输出的 logic 的值连在一起了。这是因为使用 System.out.print(distance) 语句输出 distance 变量的值时没有换行,所以,再使用 System.out.print(logic) 语句输出 logic 变量的值时,导致两个输出结果连在一起了。

在 Java 中,除了可以使用基本的 System.out.println(...) 或者 System.out.print(...) 输出数据外,还可以使用更为灵活的以下语句输出多个结果。

```
System.out.format(格式串, 数据1, 数据2, ...);
```

例如，使用以下语句同时输出三个变量的值。

```
System.out.format("count=%d, distance的值=%d, logic=%b", count, distance,
logic);
```

执行这条语句，显示如图 2-7 所示的结果。

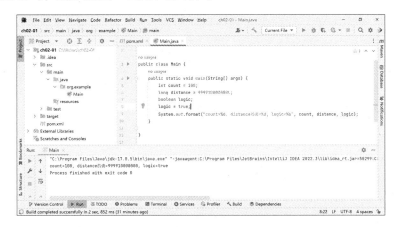

图 2-7　执行 **System.out.format** 语句的结果

这条语句的作用如图 2-8 所示。

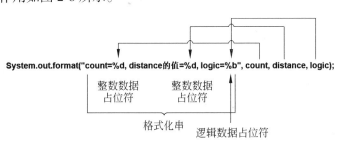

图 2-8　多数据输出语句的作用

在 System.out.format(...) 语句的小括号中有四个用英文逗号","分隔的数据，这些数据称为函数参数。在这里，第一个参数是格式化串，从字面上看，它就是一个普通字符串。这个字符串中包含类似 %d、%b 的称为占位符的特殊符号，每个占位符为后面要显示的数据占位：格式化串中有多少个占位符，在后面就应有相同数目的要显示的数据个数。Java 常用占位符及其含义如表 2-1 所示。

表 2-1　Java 常用占位符及其含义

序号	占位符	作　　用	序号	占位符	作　　用
1	%d	十进制整数占位符	5	%s	字符串占位符
2	%x	十六进制整数占位符	6	%c	字符占位符
3	%o	八进制整数占位符	7	%b	布尔值占位符
4	%f	浮点数占位符			

2.5.2 基本输入语句

程序的基本任务是对原始数据进行处理并将处理结果显示出来。前一小节介绍了如何输出（显示）数据，这一小节将介绍如何获取要处理的原始数据。一种最为简单和直接的获取要处理原始数据的方法是从键盘输入数据。Java 使用称为 Scanner 的对象从键盘接收数据，如下面的 Main.java 代码所示（源代码为 02-03.java）。

```java
package org.example;
import java.util.Scanner;
public class Main {
    public static void main(String[] args) {
        int student_in_room;
        float score_math_zhangsan;
        Scanner sc = new Scanner(System.in);
        System.out.println("请输入学生总数：");
        student_in_room = sc.nextInt();
        System.out.println("请输入张三的数学成绩：");
        score_math_zhangsan = sc.nextFloat();
        System.out.println(student_in_room);
        System.out.println(score_math_zhangsan);
    }
}
```

在这段代码中，先通过语句：

```
Scanner sc = new Scanner(System.in);
```

创建了可以从键盘进行数据输入的变量，然后使用语句：

```
student_in_room = sc.nextInt();
```

和

```
score_math_zhangsan = sc.nextFloat();
```

分别从键盘输入一个整数和一个浮点数且都保存在带响应的变量中。然后,通过以下语句：

```
System.out.println(student_in_room);
System.out.println(score_math_zhangsan);
```

将变量的值显示在屏幕上。运行这个程序，结果如图 2-9 所示。

第 2 章　Java 基本运算和输入／输出

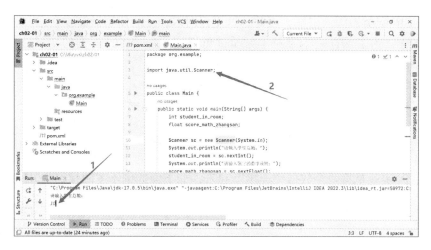

图 2-9　执行基本输入语句的结果

在图 2-9 中，在箭头 1 所指处输入数字 10 并按 Enter 键，接着输入数字 89.5 并按 Enter 键，输入这两个数后会保存到相应的变量中，并通过后续的两个输出语句显示在屏幕上。可以使用如表 2-2 所示的 Scanner 的方法输入指定类型的值到目标变量中。注意，在可以使用 Scanner 之前，需要加上箭头 2 指向的导入语句（后面章节会介绍为什么需要这条语句）。

表 2-2　Scanner 对象常用输入数据的方法

方　　法	作　　用	举　　例
int nextInt( )	从键盘读取一个 int 类型整数	Scanner sc = new Scanner(System.in); int count = sc.nextInt( );
long nextLong( )	从键盘读取一个 long 类型整数	Scanner sc = new Scanner(System.in); long distance = sc.nextLong( );
float nextFloat( )	从键盘读取一个 float 类型浮点数	Scanner sc = new Scanner(System.in); float salary = sc.nextFloat( );
double nextDouble( )	从键盘读取一个 double 类型浮点数	Scanner sc = new Scanner(System.in); double gdp = sc.nextDouble( );
boolean nextBoolean( )	从键盘读取一个 boolean 类型逻辑值	Scanner sc = new Scanner(System.in); boolean logic = sc.nextBoolean( );
String nextLine( )	从键盘读取一个 String 字符串	Scanner sc = new Scanner(System.in); String name = sc.nextString( ); // 第 2.5.3 小节将介绍字符串 String 类的使用

2.5.3　String 类的使用

字符串是程序设计中最常用的数据类型之一，如 "Hello, world!" 就是一个字符串字面常量。更多时候，需要将一个字符串字面常量保存到一个变量中，因此，需要了解 Java 的重要数据类型 String 类的使用。

String 类不是 Java 的基本数据类型，而是在 JDK（Java development kit, Java 的软件开发工具包）中定义的一个类。在深入介绍类、对象这些基本的面向对象概念之前，既不需要也不能对 String 类做深入讲解，因此，这里仅对 String 的使用做简要介绍。

在 Java 中，为了保存一个字符串常量，可以定义一个 String 类的变量，然后，将一个字符串常量赋值给这个变量，如下面的例子。

```
String name;
name = "我的名字叫张三";
```

字符串也支持"加法"运算：将两个字符串"加"起来，或者将一个字符串与数值"加"起来，其实就是将两个数据"连接"起来，如下面的例子（源代码为 02-04.java）。

```
String name = "我的名字叫张三";
int age = 20;
String some = name + "我今年" + age + "岁。";
System.out.println(some);
```

如果将这段代码放在 main( ) 函数中并运行，如图 2-10 所示，则显示"我的名字叫张三我今年 20 岁。"这段文字。

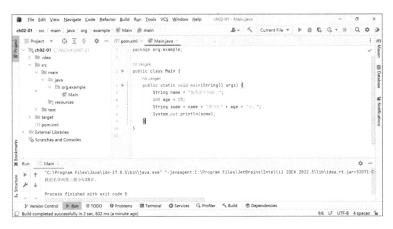

图 2-10　String 类的使用结果

2.6　练习：计算工资

编写一个工资计算程序，该程序提示用户输入工人的名字、工作天数和每天的工资，然后计算应该发放给这位工人的工资，并显示出来。程序运行过程和效果如下所示。

```
请输入工人名字: 张三
请输入工作天数: 10
请输入每天工资数额: 108.5
名字: 张三
应发工资: 1085.0
```

第 3 章　Java 程序流程控制

原则上，计算机程序都是从上到下顺序被执行的。但是，在计算机程序从上到下顺序执行的过程中，可能会因为某些条件的变化跳过某些语句的执行，或者因为某种原因需要重复执行某些语句。为了控制计算机程序执行的过程或者走向，需要使用程序设计语言中的流程控制语句。本章对流程控制语句进行介绍。

3.1　顺序语句和 if 分支语句

顺序语句是指用一对大括号"{ }"括起来的从上到下逐条顺序被执行的一组语句。分支流程控制语句包括 if 语句、if...else 语句、if...else if...else if...else 语句、switch...case default 语句等。由于 Java 对 switch...case default 分支语句做了增强，将在下一节专门介绍 switch...case default 分支语句。

顺序语句和 if 分支语句

3.2　switch...case default 分支语句及其应用实践

switch 语句也称为多分支语句。它用于检查某个变量的值，如果这个变量的值与某个 case 语句后面的值相等，则执行这个 case 后面的语句。如果这个值不等于任何 case 的值，则执行 default 后面的语句。如下面的例子所示（源代码为 03-01.java）。

```java
public static void main(String[] args) {
    Scanner sc = new Scanner(System.in);
    System.out.println("请输入一个单词: ");
    String str = sc.nextLine();
    switch (str) {
        case "hello":
            System.out.println("你说: " + "hello");
            System.out.println("我说: 你好!");
            break;
        case "world":
            System.out.println("你说: "+ "world");
            System.out.println("我说: 中国!");
            break;
```

```
        default:
            System.out.println("I don't know!");
            break;
    }
    System.out.println("Over!");
}
```

这段代码，先提示输入一个单词，程序从键盘读取这个单词并保存到 str 这个字符串变量中，然后通过 switch 语句检查 str 变量的值，若等于 hello，则执行第一个 case 子句中的语句，即显示"你说："+"hello"和"我说：你好！"后，执行 break 语句，跳出 switch 语句，然后执行 switch 大括号后面的代码，这里是执行"System.out.println("Over!");"语句，显示"Over!"；如果 str 变量中的值不等于 hello，则检查 str 中的值是否等于 world，若是，则执行第二个 case 子句中的语句，即显示"你说："+"world"和"我说：中国！"后，执行 break 语句，跳出 switch 语句，然后执行 switch 大括号后面的代码，这里是执行"System.out.println("Over!");"语句，显示"Over!"。以此类推。如果 str 不等于任何 case 子句中的值，则执行 default 子句中的语句，显示"I don't know!"后，再执行紧跟的 break 语句，跳出 switch 语句，最后继续执行"System.out.println("Over!");"语句，显示"Over!"。

上面代码中，break 语句的作用是：跳出对应的 case 子句或 default 子句，继续执行 switch 大括号后面的语句。如果某个 case 子句对应的子句中没有 break 语句，程序会继续执行后面的 case 或 default 子句中的代码，直到碰到 break 语句或者 switch 语句则执行完成。

switch...case default 语句的一般形式如下：

```
switch(变量) {
    case 常量1:
        语句;
        ...
        break;      //break 语句是可有可无的
    case 常量2:
        语句;
        ...
        break;      //break 语句是可有可无的
    case 常量3:
        语句;
        ...
        break;      //break 语句是可有可无的
    default:        //default 子句是可有可无的
        语句;
        ...
        break;      //break 语句是可有可无的
}
...
```

switch 语句有一个容易让程序员产生错误（也称为 bug）的地方：如果在某个 case 子

句中，本来需要使用 break 语句跳出 switch 语句，可能由于粗心忘记了使用 break 从而导致继续执行该 case 子句紧接着的 case 子句代码。新版 JDK 对 switch 语句的使用方法（语法）做了简化。例如，下面的例子与本节开头的例子是完全等价的（源代码为 03-02.java）。

```java
public static void main(String[] args) {
    Scanner sc = new Scanner(System.in);
    System.out.println("请输入一个单词:");
    String str = sc.nextLine();
    switch (str) {
        case "hello" -> {
            System.out.println("你说: " + "hello");
            System.out.println("我说: 你好! ");
        }
        case "world" -> {
            System.out.println("你说: " + "world");
            System.out.println("我说: 中国! ");
        }
        default -> System.out.println("I don't know!");
    }
    System.out.println("Over!");
}
```

使用"->"这种新的 switch...case 结构，Java 会在每个"->"所对应的代码末尾自动加上 break 语句，从而解决程序员容易忘记 break 语句的问题。同时需要强调一下，在 switch 语句中，要么使用"case ->"形式的分支形式，要么使用"case :"形式的分支形式，不可混用。

switch 多分支语句使用的应用实践是：用于进行 switch 比较的变量，不要使用浮点数据类型的变量。因为，对于一些浮点数据，计算机不能对其进行精确表示，因此，这种情况会导致一些不能预期的情况发生。

3.3 循环语句

有时需要重复执行一段代码。例如，为了进行从 1 到 100 的加法运算，需要不断地执行加法运算。为此，需要使用循环语句。Java 常用的循环语句包括 for 循环和 while 循环。

循环语句

3.4 数组

当程序中需要存储大量同类型的数据时，需要使用数组。例如，为了保存 50 个学生的语文考试成绩，如果没有数组，需要定义 50 个变量，这非常不方便，而如果使用数组，则只需要定义一个数组变量即可。

数组

3.4.1 定义数组

在 Java 中，定义数组需要两步：第一步，定义数组变量的名称和数据类型；第二步，确定数组的元素个数。定义数组变量的名称和数据类型的一般形式如下：

> **数据类型 [] 数组变量名；**

这里只完成了第一步，也就是说，只定义了数组的数据类型和变量名称，并没有说明这个数组有多少个元素。为了说明数组的元素个数，需要使用 Java 的 new 关键字为数组创建存储空间。确定数组元素个数的一般形式如下：

> **数组变量名 = new 数据类型 [元素个数]；**

在使用 new 关键字确定数组元素时，Java 会自动地将每个数组元素的值初始化为数值 0。例如，为了定义一个能够保存 50 个学生语文成绩的数组变量，可以使用以下语句：

```
float[] scores;
scores = new float[50];
```

在 Java 中，也可以将定义数组的两个步骤合为一个步骤，如下所示。

> **数据类型 [] 数组变量名 = new 数据类型 [元素个数]；**

类似地，为了定义一个能够保存 50 个学生语文成绩的数组变量，可以使用以下语句。

```
float[] scores = new float[50];
```

还可以在定义数组的同时完成对数组元素的初始化。

> **数据类型 [] 数组变量名 = new 数据类型 [] { 初始值 0, 初始值 1, ..., 初始值 k }；**

或者

> **数据类型 [] 数组变量名 = { 初始值 0, 初始值 1, ..., 初始值 k }；**

采用这种方式定义并初始化一个数组时，大括号里面的元素个数就是数组的元素个数。

3.4.2 访问数组元素

一旦定义了数组并通过 new 关键字确定了数组的元素个数，就可以使用"[下标]"

运算符访问数组的元素,其中"下标"是一个整数值:假定数组的元素个数为 N,那么,"下标"取值范围为 0、1、2、…、N-1,表示要访问数组的第 1、2、…、N 个元素的内容。看下面这个例子(源代码为 03-03.java)。

```java
public static void main(String[] args) {
    float[] salary = new float[12];
    salary[0] = 5000.15F;
    salary[1] = 6060.6F;
    System.out.println("1月的收入是: " + salary[0]);
    System.out.println("2月的收入是: " + salary[1]);
    System.out.println("3月的收入是: " + salary[2]);
    System.out.println("12月的收入是: " + salary[11]);
}
```

在这段代码中,首先定义一个 float 类型的数组,并确定数组的元素为 12,用于保存 12 个月的收入,然后初始化 1 月和 2 月的收入分别为 5000.15 元和 6060.6 元,最后使用输出语句显示 1 月、2 月、3 月和 12 月的收入,运行结果如下:

```
1月的收入是: 5000.15
2月的收入是: 6060.6
3月的收入是: 0.0
12月的收入是: 0.0
```

访问数组元素的一般形式如下:

数组变量名 [下标]

3.4.3 使用 for each 遍历数组元素

使用数组类似于定义了多个同名同类型的变量,数组的这一特性非常适合用于对数组的遍历:访问数组中的每一个元素。当然,为了遍历数组中的元素,可以使用前面介绍的 for 循环或者 while 循环。例如,下面的例子分别使用 for 循环和 while 循环遍历数组以计算一年的总收入(源代码为 03-04.java)。

```java
public static void main(String[] args) {
    float[] salary = {5000.0F, 1234.5F, 6090.5F, 8000F, 6781.5F, 6712.8F,
                      2312.4F, 3456.7F, 10000.5F, 2111.8F, 4070F, 8760.5F};
    float total = 0;
    for(int i=0; i<12; i++) {
        total += salary[i];
    }
    System.out.println("全年的收入是: " + total);
}
```

这个程序定义并初始化一个数组，然后使用 for 循环计算全年的总收入，或者使用 while 循环完成同样的功能（源代码为 03-05.java）。

```java
public static void main(String[] args) {
    float[] salary = {5000.0F, 1234.5F, 6090.5F, 8000F, 6781.5F, 6712.8F,
                2312.4F, 3456.7F, 10000.5F, 2111.8F, 4070F, 8760.5F};
    float total = 0;
    int i = 0;
    while(i<12) {
        total += salary[i];
        i++;
    }
    System.out.println("全年的收入是: " + total);
}
```

在 Java 中，为了遍历数组元素，还可以使用一种更为简单的方式，即使用 for each 循环，如下面的例子所示（源代码为 03-06.java）。

```java
public static void main(String[] args) {
    float[] salary = {5000.0F, 1234.5F, 6090.5F, 8000F, 6781.5F, 6712.8F,
                2312.4F, 3456.7F, 10000.5F, 2111.8F, 4070F, 8760.5F};
    float total = 0;
    for(float m : salary)
        total += m;
    System.out.println("全年的收入是: " + total);
}
```

for each 的一般形式如下：

```
for(数组元素类型 变量 : 数组变量) {
    语句1;
    语句2;
    ...
}
```

for each 循环的执行逻辑是：先将数组中的每个元素值逐一地保存到"变量"中，再执行大括号里面的语句代码。

3.4.4 二维数组

在 Java 中既可以定义一维数组，也可以定义二维数组，甚至更高维的数组。二维数组类似于日常生活中的表格：包括数据行和数据列。例如：

```
90.5    78.5    91.2    100     90
89.3    100     90.5    98.5    91.5
78.5    90.4    100     98      99.5
```

这是一个 3 行 5 列的数据。在 Java 中，为了存储类似表格的数据，可以使用二维数组。为了存储这个数据，可以使用以下语句定义数组并初始化元素的值。

```
double[][] scores = new scores[3][5];
scores[0][0] = 90.5;
scores[0][1] = 78.5;
scores[0][2] = 91.2;
scores[0][3] = 100;
scores[0][4] = 90;
scores[1][0] = 89.3;
...// 此处省略了部分代码
scores[2][0] = 78.5;
...// 此处省略了部分代码
scores[2][4] = 99.5;
```

或者，在定义数据的同时初始化数据元素的值。

```
double[][] scores = new double[][] {
    {90.5, 78.5, 91.2, 100, 90},
    {89.3, 100, 90.5, 98.5, 91.5},
    {78.5, 90.4, 100, 98, 99.5}
};
```

可以使用以下语句显示第 3 行第 5 列的数据。

```
System.out.println(scores[2][4]);
```

在 Java 中，定义二维数组的一般形式如下：

数据类型 [ ][ ] 数组名 ;

然后使用 new 关键字确定数组的大小。

数组名 = new 数据类型 [数组的行数][数组的列数];

或者，在定义数组的同时初始化数组的元素值。

数据类型 [ ][ ] 数组名 = new 数据类型 [ ][ ] {
    { 第 1 行数据，各数据之间用逗号 "," 分隔 },

```
    { 第 2 行数据，各数据之间用逗号 "," 分隔 },
    ...
};
```

访问二维数组元素的一般形式如下：

```
数组名 [ 行下标 ][ 列下标 ]
```

其中，"行下标"取值范围为从 0 到数组的行数减 1；"列下标"取值范围为从 0 到数组的列数减 1。

3.5　switch 表达式和 yield 关键字的使用

switch 表达式用于判断"switch(参数)"小括号中参数值，并结合 case 子句及"->"符号返回一个值到某个目标变量，或者结合 case 子句和 yield 语句返回一个值到某个目标变量。例如，下面这个例子，根据输入的从 1 到 7 的整数，通过 switch 表达式返回相应的星期名到目标变量中（源代码为 03-07.java）。

```java
public static void main(String[] args) {
    Scanner sc = new Scanner(System.in);
    System.out.println("请输入1到7的数字: ");
    int which = sc.nextInt();
    String name = switch (which) {
        case 1 -> "星期一";
        case 2 -> "星期二";
        case 3 -> {
            yield "星期三";
        }
        case 4 -> "星期四";
        case 5 -> {
            yield "星期五";
        }
        case 6 -> "星期六";
        case 7 -> "星期天";
        default -> "非法输入";
    };
    System.out.println(name);
}
```

执行这个程序，输入 3，将显示如图 3-1 所示的结果。

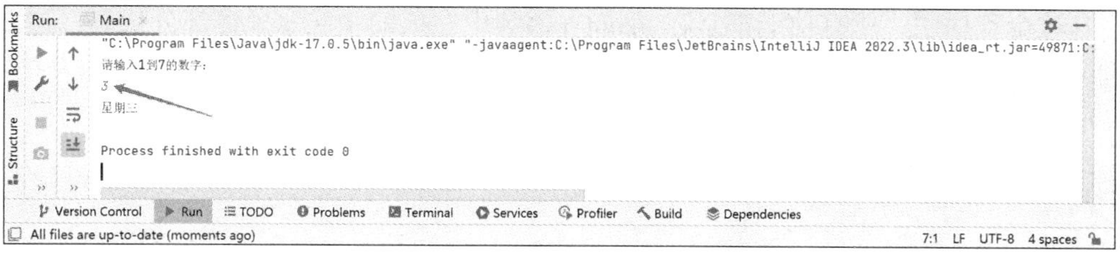

图 3-1　switch 表达式的使用结果

3.6　函数及其调用

到目前为止，Main 主类中只有一个 main( ) 函数，也就是入口函数。当程序功能复杂且代码量比较大时，可以根据业务功能将程序拆分为多个功能部件，这些功能部件称为函数，也称为方法。一个 Java 类可以包含多个函数。函数是规模化程序设计的有力手段。

下面举例子说明函数及其使用。程序不断从键盘接收用户输入的任意一个整数，然后告诉用户所输入的数是不是质数，输入 0 则结束程序。在没有使用函数这种方式的情况下，代码如下（源代码为 03-08.java）：

```java
package org.example;
import java.util.Scanner;
public class Main {
    public static void main(String[] args) {
        Scanner sc = new Scanner(System.in);
        while(true) {
            System.out.println("请输入一个正整数：");
            int n = sc.nextInt();
            if (n <= 0) break;
            boolean is_prime = true;
            for(int i=2; i<=(n/2); i++) {
                if ((n%i) == 0) {
                    is_prime = false;
                    break;
                }
            }
            if (is_prime) {
                System.out.println("所输入的数 '" + n + "' 是质数 \n");
            }
            else {
                System.out.println("所输入的数 '" + n + "' 不是质数 \n");
            }
        }
    }
}
```

在这个程序中,所有功能都是在 main( ) 函数中完成的。注意黑体部分代码,这段代码判断所输入的 n 是不是质数。如果使用函数重写这个程序,代码如下(源代码为 03-09.java):

```java
package org.example;
import java.util.Scanner;
public class Main03 {
    public static void main(String[] args) {
        Scanner sc = new Scanner(System.in);
        while(true) {
            System.out.println("请输入一个正整数: ");
            int n = sc.nextInt();
            if (n <= 0) break;
            boolean yes_or_no;
            yes_or_no = judgePrime(n);                    // 函数调用
            if (yes_or_no) {
                System.out.println("所输入的数 '" + n + "' 是质数 \n");
            }
            else {
                System.out.println("所输入的数 '" + n + "' 不是质数 \n");
            }
        }
    }
    public static boolean judgePrime(int number) {        // 函数定义
        boolean is_prime = true;
        for(int i=2; i<=(number/2); i++) {
            if ((number%i) == 0) {
                is_prime = false;
                break;
            }
        }
        return is_prime;
    }
}
```

在这个程序中有两个函数:一个是 main( ) 入口函数,另一个是 judgePrime( ) 函数。注意其中的语句。

```java
public static boolean judgePrime(int number) {        // 函数定义
```

这条语句定义了一个函数:函数的名字为 judgePrime,它需要一个 int 类型的参数,并且该函数会返回一个 boolean 类型的值。例如,在 main( ) 函数中使用以下语句调用这个函数时:

```
yes_or_no = judgePrime(n);        // 函数调用
```

Java 虚拟机会将 n 的值（n 称为实参）复制到 number 参数中（number 称为形参）中，然后执行 judgePrime 函数的代码。

```
boolean is_prime = true;
for(int i=2; i<=(number/2); i++) {
    if ((number%i) == 0) {
        is_prime = false;
        break;
    }
}
return is_prime;
```

这段代码将判断参数 number（这里就是实参 n 的值）是否为质数，若是，则返回 true；否则返回 false。因此，从形式上可以理解为函数名字 judgePrime( ) 指代了以上一段代码。当函数执行完成并返回时，所返回的值将被保存到 yes_or_no 这个变量中，进而，main( ) 函数的代码可以得到 judgePrime( ) 函数的执行结果，并根据结果显示所输入的数是否为质数。

在后续章节中将介绍函数定义中所涉及的 public static 关键字的作用，此时可以不理会这两个关键字的含义及作用。

3.7 案例：学生成绩计算系统

下面通过一个例子，对本章及本章之前的知识进行综合应用，并结束本章的内容。

3.7.1 案例任务

全班有 20 个学生，近期进行了数学课阶段考试，需要编写一个 Java 程序对成绩进行统计。编写一个程序，首先输入全班的考试成绩，然后计算并输出学生成绩的最高分、最低分和平均分。

3.7.2 任务分析

要输入和保存 20 个学生的数学成绩，需要定义一个 float 类型的数组变量，并指定数组的元素个数为 20，当然，也需要定义保存最高分、最低分和平均分的变量。之后，使用 for 循环或者 while 循环提示输入成绩，在输入成绩的同时计算成绩的最高分、最低分和平均分。

3.7.3 任务实施

Main 类代码如下（源代码为 03-10.java）：

```java
package org.example;
import java.util.Scanner;
public class Main {
    public static void main(String[] args) {
        Scanner sc = new Scanner(System.in);
        float[] scores = new float[20];
        float max = -10.F, min = 200.0F, avg = 0.0F;
        for(int i=0; i<20; i++) {
            System.out.println("请输入第" + i + "个学生的成绩：");
            scores[i] = sc.nextFloat();
            if (max < scores[i])
                max = scores[i];
            if (min > scores[i])
                min = scores[i];
            avg = avg + scores[i];
        }
        avg = avg / 20;
        System.out.println("最高分：" + max);
        System.out.println("最低分：" + min);
        System.out.println("平均分：" + avg);
    }
}
```

这里用到了一个编程技巧：利用一个循环即可计算最高分、最低分和平均分。还需要用到一个常识：最高成绩不可能为负数，最低成绩也不可能是 200 分，所以，一旦输入一个成绩后，max 和 min 变量的值将回归正确的值。

3.8 练习：计算质数及其和

编写一个 Java 程序，计算从 1 到任意整数之间的所有质数及其和。程序首先要求输入一个正整数 n，然后计算并输出从 1 到正整数 n 之间的所有质数及其和。

第 4 章 类 和 对 象

前面介绍了 Java 作为一门程序设计语言的基本内容。从本章开始，将介绍 Java 的重要内容——面向对象编程。面向对象不仅是一种编程技术，更是一种编程思想。从面向对象视角，任何程序都是为了达成预定功能目标而由多个相互协同、相互作用的对象构成的整体。Java 语言为这种编程思想提供了技术实现手段。使用 Java 语言，结合面向对象编程思想，可以编写出功能强大且形式优雅的程序。

4.1 定义类和创建对象

程序是由对象构成的。那么，什么是对象？对象又是如何产生的？对象之间又是如何相互协同达成程序功能目标的？要了解什么是对象，必须先了解类。

4.1.1 类的含义

所谓"物以类聚"，就是指根据物品的种类不同进行分门别类的管理。观察一下如图 4-1 所示的物品归类示意图。

图 4-1 物品归类示意图

任何一个观察者可以毫不含糊地将图 4-1 中的物品描述为：左边是汽车，中间是建筑物，右边是人员。如果仔细分析这段描述会发现：这段描述抓住了事物的本质。例如，对于"左边是汽车"这段描述，虽然这些汽车的颜色不同、品牌不同、类别不同、使用能源不同，但是，它们都具有汽车这个种类所具有的共同特征：有轮子，可以驾驶。类似地，对于"中间是建筑物"这段描述，虽然这两个建筑物的类别不同、高度不同、风格不同，但是它们都具有建筑物这个种类所具有的共同特征：可以遮风挡雨，可以住人，可以办公。对于"右边是人员"这段描述的分析类似，无须赘述。

在 Java 语言中，使用类来描述同类事物的本质特征。使用类描述事物本质特征的过程称为定义类。

4.1.2 定义类

在 Java 中，使用关键字 class 来定义同类事物的本质特征。例如，可以采用以下代码定义图 4-1 左边的汽车类（源代码为 04-01.java）。

```java
public class Vehicle {
    String name;
    String manufacture;
    int wheels;
    int seats;
    int fuelType;     //1代表Gasoline, 2代表Electricity
    public void info() {
        System.out.println("Name: " + name);
        System.out.println("Manufacture: " + manufacture);
        System.out.println("Wheels: " + wheels);
        System.out.println("Seats: " + seats);
        String t = (fuelType==1?"Gasoline":"Electricity");
        System.out.println("Fuel Type: " + t);
    }
}
```

上面这段代码中，使用：

```java
public class Vehicle {
...
}
```

定义了一个名为 Vehicle 的 Java 类。处于一对大括号 "{ }" 之间的代码是对 Vehicle 类的具体描述。其中：

```java
String name;
String manufacture;
int wheels;
int seats;
int fuelType;     //1代表Gasoline, 2代表Electricity
```

用于描述 Vehicle 类的属性。试想一下，任何一台汽车都具有以下属性：名字、生产商、轮子数、可容纳人数、所使用的能源类型等。当然，由于程序的设计目标不同，即使为同一种物体定义 Java 类也可能是不相同的，也就是说，在设计一个类时，应该根据程序的设计目标确定类应该包含的属性。

在类的定义中，除了可以包含属性外，还可以包含方法（也称为函数）。例如，在汽车这个类中，定义了以下方法。

```java
public void info() {
    System.out.println("Name: " + name);
    System.out.println("Manufacture: " + manufacture);
    System.out.println("Wheels: " + wheels);
    System.out.println("Seats: " + seats);
    String t = (fuelType==1?"Gasoline":"Electricity");
    System.out.println("Fuel Type: " + t);
}
```

这个方法显示一台汽车的相关信息，包括铭牌、生产商、轮子数、可容纳人数和所使用的能源类型。

从上面这个 Vehicle 类的定义可以看出，类是对同类型事物的共性抽象。这些共性既可以是属性，也可以是方法，其中，类的属性用于描述类的特征，类的方法用于定义类的行为。

为了对定义类的过程和步骤有进一步的理解，再定义一个学生类 Student。

首先分析学生这个群体所具有的共同属性。通过分析发现，任何一个学生都具有如下属性：姓名、学号、联系电话、家庭地址。再者，根据程序设计目的需要，再分析一下学生这个群体所具有的行为。通过分析，这些行为应该包括：显示学生信息和录入学生信息。为此，定义 Student 类的 Java 代码如下（源代码为 04-02.java）：

```java
public class Student {
    String name;
    String sid;
    String phone;
    String address;
    public void display() {
        System.out.println("姓名:" + name);
        System.out.println("学号:" + sid);
        System.out.println("联系电话:" + phone);
        System.out.println("家庭地址:" + address);
    }
    public void input_info() {
        Scanner sc = new Scanner(System.in);
        System.out.println("请输入学生姓名:");
        name = sc.nextLine();
        System.out.println("请输入学生学号:");
        sid = sc.nextLine();
        System.out.println("请输入学生电话:");
        phone = sc.nextLine();
        System.out.println("请输入学生地址:");
```

```
        address = sc.nextLine();
    }
}
```

所定义的 Student 类包括四个属性和两个方法。一般地，在 Java 中定义一个 Java 类的一般格式如下：

```
public class 类名 {
    // 类的属性定义
    // 类的方法定义
}
```

其中，"类的属性定义"用于描述类的特征，其一般格式如下：

访问限定符 属性类型 属性名称 ；

可以省略属性定义中的"访问限定符"，关于访问限定符的使用将在后续章节介绍。"类的方法定义"用于描述类的行为，其一般格式如下：

访问限定符 返回值类型 方法名 (参数类型 1 参数 1，参数类型 2 参数 2，...) {
    // 方法代码
}

与"类的属性定义"类似，可以省略属性定义中的"访问限定符"。方法中的参数列表"参数类型 1 参数 1，参数类型 2 参数 2, ..."是可选的，根据不同的方法确定不同的参数类型和参数数目。

4.1.3　在 IDEA 中创建 Java 类

为了在 IDEA 中创建 Java 类，首先在 IDEA 中，依据第 1 章所介绍的步骤新建一个名为 ch04-01 的 Java 工程，如图 4-2 所示。

图 4-2　新建的 ch04-01 工程

在 Java 工程中，src 文件夹是工程总文件夹。在 src 文件夹下有两个子文件夹 main 和 test，其中，test 文件夹保存测试有关的代码和资源，main 文件夹是主要的工作文件夹。

main 文件夹下有两个子文件夹 java 和 resources。java 文件夹主要用于保存 Java 程序包、Java 源代码，而 resources 文件夹则主要用于保存程序用到的资源相关文件。例如图 4-2 中，在 java 文件夹下有一个名为 org.example 的程序包，在 org.example 包下有一个名为 Main 的类，双击 Main，即可在右边的编辑栏编辑 Main.java 类的源代码，如图 4-3 所示。

图 4-3　Main.java 类的源代码

现在，在 org.example 包下创建名为 Vehicle 的程序类，创建过程和步骤如图 4-4 所示，选择 org.example → New → Java Class 命令。

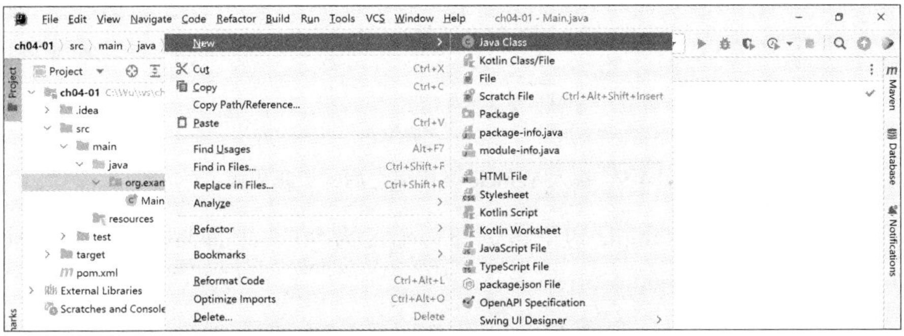

图 4-4　创建名为 Vehicle 的程序类

在弹出的输入框中输入 Vehicle，然后按 Enter 键，IDEA 将在 org.example 包下新建一个 Vehicle 程序类，并在代码编辑器中打开这个类代码等待编辑。在代码编辑器中录入 Vehicle 类的代码（源代码为 04-03.java）。

```java
package org.example;
public class Vehicle {
    String name;
    String manufacture;
    int wheels;
    int seats;
    int fuelType;   //1 代表 Gasoline，2 代表 Electricity
    public void info() {
```

```
        System.out.println("Name: " + name);
        System.out.println("Manufacture: " + manufacture);
        System.out.println("Wheels: " + wheels);
        System.out.println("Seats: " + seats);
        String t = fuelType==1?"Gasoline":"Electricity";
        System.out.println("Fuel Type: " + t);
    }
}
```

输入 Vehicle 类代码之后的工程界面如图 4-5 所示。

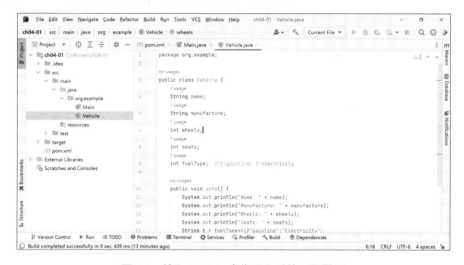

图 4-5　输入 Vehicle 类代码之后的工程界面

类似地，可以在 org.example 包下再创建一个名为 Student 的 Java 程序类。Student 类的代码如下（源代码为 04-04.java）：

```
package org.example;
import java.util.Scanner;
public class Student {
    String name;
    String sid;
    String phone;
    String address;
    public void display() {
        System.out.println("姓名: " + name);
        System.out.println("学号: " + sid);
        System.out.println("联系电话: " + phone);
        System.out.println("家庭地址: " + address);
    }
    public void input_info() {
        Scanner sc = new Scanner(System.in);
```

```
        System.out.println("请输入学生姓名：");
        name = sc.nextLine();
        System.out.println("请输入学生学号：");
        sid = sc.nextLine();
        System.out.println("请输入学生联系电话：");
        phone = sc.nextLine();
        System.out.println("请输入学生家庭地址：");
        address = sc.nextLine();
    }
}
```

创建 Student 类后的工程结构如图 4-6 所示。

图 4-6　创建 Student 类后的工程结构

从图 4-6 的工程结构可以看出，目前的 ch04-01 工程中包含一个名为 org.example 的程序包。在 org.example 包下，有一个名为 Main 的 Java 主类、一个名为 Vehicle 的程序类和一个名为 Student 的程序类。

在 IDEA 中创建 Java 类

4.1.4　创建及使用对象

现在已在 ch04-01 程序工程下创建了两个新的 Java 程序类 Student 和 Vehicle，如何在程序中使用这两个新定义的类呢？这可以基于已经定义的类创建对象并使用对象。例如，打开 Main.java 主类，修改 Main 类（源代码为 04-05.java）。

```
package org.example;
public class Main {
    public static void main(String[] args) {
        Vehicle vehicle;
        vehicle = new Vehicle();
```

```
            vehicle.name = "Audi Q5";
            vehicle.manufacture = "第一汽车公司";
            vehicle.wheels = 4;
            vehicle.seats = 5;
            vehicle.fuelType = 1;    //1 代表 Gasoline, 2 代表 Electricity
            vehicle.info();
        }
    }
```

在 Main 类的 main( ) 函数中，首先使用语句：

```
Vehicle vehicle;
```

定义了 Vehicle 类的变量 vehicle，然后使用语句：

```
vehicle = new Vehicle();
```

创建了 Vehicle 类的一个对象并使用变量 vehicle 引用这个对象。在 Java 中，使用 new 关键字从一个类创建该类的一个对象。一旦创建了 Vehicle 的对象，并通过变量 vehicle 引用了这个对象，就可以通过这个变量 vehicle 来访问和使用这个对象。例如，下面的语句就使用了 vehicle 变量设置这个新创建的 Vehicle 对象的属性值。

```
vehicle.name = "Audi Q5";
vehicle.manufacture = "第一汽车公司";
vehicle.wheels = 4;
vehicle.seats = 5;
vehicle.fuelType = 1;    //1 代表 Gasoline, 2 代表 Electricity
```

这段代码将新创建的 Vehicle 对象的 name 属性设置为 Audi Q5，将 manufacture 属性设置为"第一汽车公司"，将 wheels 属性设置为 4，将 seats 属性设置为 5，将 fuelType 属性设置为 1。最后通过语句：

```
vehicle.info();
```

也就是通过调用 vehicle 对象的 info( ) 方法显示 vehicle 对象的信息。运行这个程序，将显示以下信息。

```
Name: Audi Q5
Manufacture: 第一汽车公司
Wheels: 4
Seats: 5
Fuel Type: Gasoline
```

基于已经定义的类，可以创建多个不同的对象。例如，使用下面修改 Main 类的 main( ) 方法再次创建一个 Vehicle 类的新对象（源代码为 04-06.java）。

```
package org.example;
public class Main {
    public static void main(String[] args) {
        Vehicle vehicle;
        vehicle = new Vehicle();
        vehicle.name = "Audi Q5";
        vehicle.manufacture = "第一汽车公司";
        vehicle.wheels = 4;
        vehicle.seats = 5;
        vehicle.fuelType = 1;    //1代表Gasoline, 2代表Electricity
        vehicle.info();
        Vehicle vehi = null;     // 定义变量并使之引用空对象
        vehi = new Vehicle();
        vehi.name = "比亚迪宋";
        vehi.manufacture = "比亚迪";
        vehi.wheels = 4;
        vehi.seats = 5;
        vehi.fuelType = 2;
        vehi.info();
    }
}
```

如加黑的代码所示，此时，使用语句：

```
Vehicle vehi = null;    // 定义变量并使之引用空对象
```

定义了变量 vehi 并使之引用到 null 空对象。所谓"空对象"，就是还没有任何对象。null 称为空对象或空指针。然后这段代码又创建了 Vehicle 类的一个新对象，设置这个对象的相关属性，最后，调用该对象的 info( ) 方法显示相关信息。

除了可以创建 Vehicle 类的对象外，由于已经定义了 Student 类，同样可以根据需要创建 Student 类的对象。例如，修改 Main 类的 main( ) 函数，创建 Student 类的对象（源代码为 04-07.java）。

```
package org.example;
public class Main {
    public static void main(String[] args) {
        Vehicle vehicle;
        vehicle = new Vehicle();
        vehicle.name = "Audi Q5";
        vehicle.manufacture = "第一汽车公司";
```

```
            vehicle.wheels = 4;
            vehicle.seats = 5;
            vehicle.fuelType = 1;    //1代表Gasoline,2代表Electricity
            vehicle.info();
            Vehicle vehi;
            vehi = new Vehicle();
            vehi.name = " 比亚迪宋 ";
            vehi.manufacture = " 比亚迪 ";
            vehi.wheels = 4;
            vehi.seats = 5;
            vehi.fuelType = 2;
            vehi.info();
            Student stu = new Student();
            stu.input_info();
            stu.display();
    }
}
```

在这段代码中,使用语句:

```
Student stu = new Student();
```

创建Student类的一个新对象,并使用变量stu引用这个对象,然后使用语句:

```
stu.input_info();
```

调用stu对象的input_info()方法输入信息到使用stu对象的相应属性中,最后使用语句:

```
stu.display();
```

显示stu对象的相关信息。运行这个程序,结果如下:

```
Name: Audi Q5
Manufacture: 第一汽车公司
Wheels: 4
Seats: 5
Fuel Type: Gasoline
Name: 比亚迪宋
Manufacture: 比亚迪
Wheels: 4
Seats: 5
Fuel Type: Electricity
请输入学生姓名:
```

张三
请输入学生学号：
202200001234
请输入学生电话：
1380013××××
请输入学生地址：
广州市
姓名：张三
学号：202200001234
联系电话：1380013××××
家庭地址：广州市

创建及使用对象

4.2 构造函数

在 Java 中，一旦定义了一个类，就可以基于这个类使用 new 关键字创建类的对象，就像在第 4.1 节中创建 Vehicle 和 Student 类的对象一样。为了更为精确地创建类的对象，Java 提供了构造函数。

4.2.1 类的构造函数

在定义 Java 类时，可以定义一类特殊的函数，这类特殊函数的名字与类的名称相同并且没有函数返回值。这类特殊的函数称为构造函数。通过定义构造函数，可以在创建 Java 类的对象时，设置对象的属性值，或者执行某些特定的操作。例如，可以修改前面定义的 Vehicle 类，为其增加构造函数（源代码为 04-08.java）。

```java
package org.example;
public class Vehicle {
    String name;
    String manufacture;
    int wheels;
    int seats;
    int fuelType;   //1 代表 Gasoline，2 代表 Electricity
    public Vehicle(String name, String manufacture, int wheels, int seats, int fuelType) {
        this.name = name;
        this.manufacture = manufacture;
        this.wheels = wheels;
        this.seats = seats;
        this.fuelType = fuelType;
    }
    public void info() {
```

```
        System.out.println("Name: " + name);
        System.out.println("Manufacture: " + manufacture);
        System.out.println("Wheels: " + wheels);
        System.out.println("Seats: " + seats);
        String t = fuelType==1?"Gasoline":"Electricity";
        System.out.println("Fuel Type: " + t);
    }
}
```

在新的 Vehicle 类的定义中增加了一个构造函数，这个构造函数的定义如下：

```
public Vehicle(String name, String manufacture, int wheels, int seats, int fuelType) {
    this.name = name;
    this.manufacture = manufacture;
    this.wheels = wheels;
    this.seats = seats;
    this.fuelType = fuelType;
}
```

通过构造函数，可以在构造类的新对象时，直接指定对象的属性值。注意其中的 this 关键字的使用：在这里，由于构造函数的参数名与类的属性名称相同，因此，在构造函数中将参数值赋值给属性值时，使用了 this 关键字进行限定，其含义就是，将构造参数的参数值赋值给相应的对象属性。一旦定义了这个构造函数，就可以在 Main 类的 main( ) 函数中使用以下语句构造 Vehicle 类的新对象（源代码为 04-09.java）。

```
package org.example;
public class Main {
    public static void main(String[] args) {
        Vehicle veh = new Vehicle("Benz S320", "Merchandise", 4, 5, 1);
        veh.info();
    }
}
```

运行这个程序，将显示以下结果。

```
Name: Benz S320
Manufacture: Merchandise
Wheels: 4
Seats: 5
Fuel Type: Gasoline
```

在定义类时，可以不定义（不提供）构造函数，在这种情况下，Java 会为所定义的类提供一个默认的无参构造函数，但是，一旦在代码中定义了构造函数，Java 将不再提供无参构造函数。如果需要使用多个不同参数的构造函数，则需要使用第 4.2.2 小节介绍的构造函数重载定义自己的无参构造函数。

4.2.2 构造函数重载

　　类的构造函数可以有多个：函数名称相同，但是，函数的参数类型不同或者参数个数不同的同名函数。在面向对象中，这种现象称为函数重载。如果一个类具有多个同名的构造函数，称为构造函数重载。构造函数重载是函数重载的一种特殊情况。例如，对于 Vehicle 这个类，可以定义多个构造函数（源代码为 04-10.java）。

```java
package org.example;
public class Vehicle {
    String name;
    String manufacture;
    int wheels;
    int seats;
    int fuelType;    //1 代表 Gasoline, 2 代表 Electricity
    public Vehicle() {
        this.name = "Unknown";
        this.manufacture = "Unknown";
        this.wheels = -1;
        this.seats = -1;
        this.fuelType = 1;
    }
    public Vehicle(String name, String manufacture, int wheels, int seats, int fuelType) {
        this.name = name;
        this.manufacture = manufacture;
        this.wheels = wheels;
        this.seats = seats;
        this.fuelType = fuelType;
    }
    public void info() {
        System.out.println("Name: " + name);
        System.out.println("Manufacture: " + manufacture);
        System.out.println("Wheels: " + wheels);
        System.out.println("Seats: " + seats);
        String t = fuelType==1?"Gasoline":"Electricity";
        System.out.println("Fuel Type: " + t);
    }
}
```

　　类似地，可以为 Student 类定义构造函数。在这里，定义了三个构造函数。可用于在不同情况下构造不同的 Student 对象（源代码为 04-11.java）。

```java
package org.example;
import java.util.Scanner;
public class Student {
    String name;
    String sid;
    String phone;
    String address;
    public Student() {
        this.name = "Unknown";
        this.sid = "000000000000";
        this.phone = "000000000000";
        this.address = "Unknown";
    }
    public Student(String name) {
        this.name = name;
        this.sid = "000000000000";
        this.phone = "000000000000";
        this.address = "Unknown";
    }
    public Student(String name, String sid, String phone, String address) {
        this.name = name;
        this.sid = sid;
        this.phone = phone;
        this.address = address;
    }
    public void display() {
        System.out.println("姓名：" + name);
        System.out.println("学号：" + sid);
        System.out.println("联系电话：" + phone);
        System.out.println("家庭地址：" + address);
    }
    public void input_info() {
        Scanner sc = new Scanner(System.in);
        System.out.println("请输入学生姓名：");
        name = sc.nextLine();
        System.out.println("请输入学生学号：");
        sid = sc.nextLine();
        System.out.println("请输入学生联系电话：");
        phone = sc.nextLine();
        System.out.println("请输入学生家庭地址：");
        address = sc.nextLine();
    }
}
```

现在，可以使用 Student 类的不同构造函数和 new 关键字构造不同的对象。例如，在 Main 类的 main( ) 函数中构造三个不同的 Student 对象（源代码为 04-12.java）。

```java
package org.example;
public class Main {
    public static void main(String[] args) {
        Student stu01 = new Student();
        stu01.display();
        Student stu02 = new Student("张三");
        stu02.input_info();
        stu02.display();
        Student stu03 = new Student("李四", "202200001234", "1380013××××", "广州市");
        stu03.display();
    }
}
```

运行这个程序，显示以下结果。

```
姓名:Unknown
学号:000000000000
联系电话:000000000000
家庭地址:Unknown
请输入学生姓名:
Wanger
请输入学生学号:
202300004321
请输入学生电话:
13500135××××
请输入学生地址:
Haizhu
姓名:Wanger
学号:202300004321
联系电话:13500135××××
家庭地址:Haizhu
姓名:李四
学号:202200001234
联系电话:1380013××××
家庭地址:广州市
```

4.3 类的静态属性、静态方法和静态代码块

一旦定义了一个 Java 类并基于 Java 类使用 new 关键字创建该类的对象，所创建的每个对象都具有自己独立的属性值。例如，对于 Student 类，使用以下语句创建两个 Student

对象。

```
Student stu01 = new Student(" 张 三 ", "202300000000", "1350013××××",
" 北京市 ");
Student stu02 = new Student(" 李 四 ", "202200001234", "1380013××××",
" 广州市 ");
```

那么，对于 stu01 对象和 stu02 对象，虽然它们都有 name、sid、phone 和 address 属性，然而它们的属性值都是互不相同的。因此，当使用 stu01.display( ) 和 stu02.display( ) 方法显示对象的信息时将显示不同的值。但是，在某些特殊情况下，可能需要一个类的所有对象的某个或者某些属性具有相同的值，这时，就需要使用到静态属性。

4.3.1 静态属性

所谓静态属性，是指使用 static 关键字修饰的属性。一旦使用了 static 关键字修饰了某个类的某个属性，则这个属性值将被这个类的所有对象所共享，也就是说，该类的所有对象的这个属性值都是相同的。下面通过一个例子说明这种情况。在 ch04-01 工程下的 org.example 包下新建 Circle 类，这个类定义数学中的"圆"这个形状。Circle 类的代码如下（源代码为 04-13.java）。

```java
package org.example;
public class Circle {
    static float PI = 3.1415926f;
    float r;
    public Circle() {
        this.r = 0;
    }
    public Circle(float r) {
        this.r = r;
    }
    public float area() {
        return PI*r*r;
    }
    public float circumference() {
        return 2*PI*r;
    }
    public void info() {
        System.out.println(" 圆的半径: " + r);
        System.out.println(" 圆的面积: " + area());
        System.out.println(" 圆的周长: " + circumference());
    }
}
```

对于圆，圆的半径已经确定了圆的大小，因此，在 Circle 这个类中，定义了属性 r 以表示圆的大小。同时，为了计算圆的面积、圆的周长等数学性质，需要用到圆周率这个数学常量。因此，在 Circle 类中，使用以下语句定义圆周率。

```
static float PI = 3.1415926f;
```

使用 static 关键字对 PI 这个属性进行修饰，表示所有 Circle 类的对象都将共享 PI 这个属性。在计算圆的周长和面积的方法中都使用了 PI 这个属性值。

现在修改 Main 类的代码用以创建 Circle 类的对象，并显示相应圆的信息，修改后的 Main 类代码如下（源代码为 04-14.java）：

```
package org.example;
public class Main {
    public static void main(String[] args) {
        Circle c = new Circle(10.0f);
        c.info();
        Circle c2 = new Circle(20.0f);
        c2.info();
        System.out.println(c.PI);
        System.out.println(c2.PI);
        System.out.println(Circle.PI);
    }
}
```

在 main( ) 方法中创建了两个半径分别为 10.0 和 20.0 的圆对象，然后分别显示了圆的信息。注意以下用于显示 PI 属性值的代码。

```
System.out.println(c.PI);
System.out.println(c2.PI);
System.out.println(Circle.PI);
```

这段代码说明，可以使用对象 c 和 c2 来访问属性 PI 的值。但是，由于属性 PI 被 static 关键字修饰，所以，PI 属性属于 Circle 这个类，因此，可以直接使用"类名.静态属性"的方式访问 PI 的属性值，并且这种访问静态属性的方式是 Java 建议的规范方式。运行这个程序，显示以下结果。

```
圆的半径：10.0
圆的面积：314.15924
圆的周长：62.83185
圆的半径：20.0
圆的面积：1256.637
圆的周长：125.6637
```

```
3.1415925
3.1415925
3.1415925
```

4.3.2 静态方法

与类的静态属性类似,类也可以具有静态方法。静态方法是指使用 static 关键字修饰的方法。一旦使用 static 关键字修饰了类的某个方法,则这个方法属于类,可以直接使用"类名.静态方法"方式访问静态方法。下面修改第 4.3.1 小节定义的 Circle 这个类,增加一个用于修改 PI 这个静态属性的方法。修改后的 Circle 类的代码如下(源代码为 04-15.java):

```java
package org.example;
public class Circle {
    static float PI = 3.1415926f;
    float r;
    public Circle() {
        this.r = 0;
    }
    public Circle(float r) {
        this.r = r;
    }
    public float area() {
        return PI*r*r;
    }
    public static void setPI(float pi) {
        PI = pi;
    }
    public float circumference() {
        return 2*PI*r;
    }
    public void info() {
        System.out.println("圆的半径：" + r);
        System.out.println("圆的面积：" + area());
        System.out.println("圆的周长：" + circumference());
    }
}
```

在这个类中,新增加了静态方法 setPI 方法,用于修改静态属性 PI 的值。

```java
public static void setPI(float pi) {
    PI = pi;
}
```

现在，修改 Main 类的代码，在计算圆的周长和面积之前，调用 setPI 方法修改属性 PI 的值。修改后的 Main 类代码如下（源代码为 04-16.java）：

```java
package org.example;
public class Main {
    public static void main(String[] args) {
        Circle.setPI(3.14f);
        Circle c = new Circle(10.0f);
        c.info();
        Circle c2 = new Circle(20.0f);
        c2.info();
        System.out.println(c.PI);
        System.out.println(c2.PI);
        System.out.println(Circle.PI);
    }
}
```

注意语句：

```java
Circle.setPI(3.14f);
```

这条语句修改了静态属性 PI 的值。运行这个程序，显示以下结果。

```
圆的半径: 10.0
圆的面积: 314.0
圆的周长: 62.800003
圆的半径: 20.0
圆的面积: 1256.0
圆的周长: 125.600006
3.14
3.14
3.14
```

4.3.3 静态代码块

静态代码块是指被 static 关键字修饰的一段代码。当 Java 虚拟机加载被 static 关键字修饰的类代码时，Java 虚拟机会自动执行这些被 static 关键字修饰的代码块。例如，如果在 Circle 类中，希望在 Java 虚拟机完成对 Circle 类的加载后马上执行一段代码，则可以使用 static 关键字对这段代码进行修饰。修改后的 Circle 代码如下（源代码为 04-17.java）：

```java
package org.example;
public class Circle {
```

```java
    static float PI = 3.1415926f;
    float r;
    static {
        System.out.println("这是Circle的static代码块中的代码");
        System.out.println("被static关键字修饰的代码块被首先执行");
    }
    public Circle() {
        this.r = 0;
    }
    public Circle(float r) {
        this.r = r;
    }
    public float area() {
        return PI*r*r;
    }
    public static void setPI(float pi) {
        PI = pi;
    }
    public float circumference() {
        return 2*PI*r;
    }
    public void info() {
        System.out.println("圆的半径: " + r);
        System.out.println("圆的面积: " + area());
        System.out.println("圆的周长: " + circumference());
    }
}
```

注意其中被 static 关键字修饰的代码。

```java
static {
    System.out.println("这是Circle的static代码块中的代码");
    System.out.println("被static关键字修饰的代码块被首先执行");
}
```

这段代码将在 Java 虚拟机完成 Circle 类的加载时被执行。不对 Main 类做任何修改。执行这个程序，显示以下结果。

```
这是Circle的static代码块中的代码
被static关键字修饰的代码块被首先执行
圆的半径: 10.0
圆的面积: 314.0
圆的周长: 62.800003
圆的半径: 20.0
圆的面积: 1256.0
```

```
圆的周长：125.600006
3.14
3.14
3.14
```

4.3.4 静态属性、静态方法和静态代码块应用实践

在定义类时，如果一个属性是这个类的所有对象所共享的，那么，就将这个属性定义为静态属性；否则，定义这个属性为非静态属性。相应地，如果程序在不同的场景下需要对静态属性进行修改，那么，需要定义用于修改静态属性的静态方法。在定义 Circle 这个类已经清楚地例示了这种用法。

关于静态代码块的应用实践，则是主要用于对类的数据或状态进行初始化操作。一个常见的例子就是单实例（singleton）对象的使用。关于这方面的例子会在后续章节中进行介绍。

4.4　内　部　类

之前定义的 Java 类，包括 Student、Vehicle、Circle 类，都是在 org.example 包下独立的与类名同名的 Java 源文件中定义的，并且每个类的定义都使用了类似 public class ×××的形式。例如，对于 Student 类，其定义形式如下：

```
public class Student {
    // 属性定义
    // 方法定义
}
```

Vehicle 类和 Circle 类的定义也类似，这样的类称为外部公共类。

因为这些外部公共类都是在独立的 Java 源代码文件中定义的，并且都具有 public 访问限定符（关于访问限定符的含义和使用，将在第 5 章进行介绍），这些类一旦被定义，就可以被其他类使用。例如，可以在 Main 类中通过 new 运算符创建这些类的对象。

但是，在某些场景也可以存在这样的情况：某个类可能仅在指定的另外一个外部公共类中被使用。在这种情况下，就需要使用到 Java 的内部类。所谓内部类，就是指在一个类的内存中定义的另外一个类。Java 内部类主要包括：成员内部类、局部内部类、静态内部类和匿名内部类。局部内部类使用的场景不多，在此不做介绍；匿名内部类一般与接口配合使用，因此，关于匿名内部类的使用将在第 6 章进行介绍。

4.4.1 成员内部类

成员内部类，顾名思义就是将一个类定义为另一个类的成员。为了便于叙述，将包含

内部类的类称为外部类，而将被包含的作为成员的类称为内部类。在成员内部类中，内部类可以无条件访问外部类的所有成员属性和成员方法，而外部类为了能够访问内部内的成员属性或方法，需要首先使用 new 关键字实例化内部类的对象。

下面举一个例子。定义一个 Company 类用于描述公司的信息，包括 4 个属性，其中 name 表示公司名称，address 表示公司地址，id 表示公司纳税人号码，date 表示公司成立日期，包括 info( ) 方法，用于显示公司信息。在 ch04-01 工程的 org.example 包下创建 Company 类，Company 类的定义如下（源代码为 04-18.java）：

```java
package org.example;
public class Company {
    String name;
    String address;
    String id;
    FoundDate date;
    public Company(String name, String address, String id, int year, int month, int day) {
        this.name = name;
        this.address = address;
        this.id = id;
        this.date = new FoundDate(year, month, day);
    }
    public void info() {
        System.out.println("公司名称：" + name);
        System.out.println("公司地址：" + address);
        System.out.println("公司纳税人号码：" + id);
        System.out.println("公司成立日期：" + date.toString());
    }
    class FoundDate {
        int year;
        int month;
        int day;
        public FoundDate(int year, int month, int day) {
            this.year = year;
            this.month = month;
            this.day = day;
        }
        public String toString() {
            return year + "-" + month + "-" + day;
        }
    }
}
```

在 Company 类的定义中，包括一个成员内部类 FoundDate 类，也就是说，FoundDate 类是作为 Company 类的成员而存在的。

```
class FoundDate {
    int year;
    int month;
    int day;
    public FoundDate(int year, int month, int day) {
        this.year = year;
        this.month = month;
        this.day = day;
    }
    public String toString() {
        return year + "-" + month + "-" + day;
    }
}
```

这个类用于表示公司成立的日期。同时，在 Company 类中，有一个名为 date 且类型为 FoundDate 的属性。

```
FoundDate date;
```

在 Company 的构造函数中，根据构造函数参数中的 year、month、day 参数，创建了 FoundDate 类的对象，并将对象保存到 Company 类的 date 属性中。

```
this.date = new FoundDate(year, month, day);
```

现在，修改 Main 类，在 main( ) 方法中创建 Company 类的对象，然后调用 Company 对象的 info( ) 方法显示公司信息。修改后的 Main 类代码如下（源代码为 04-19.java）：

```
package org.example;
public class Main {
    public static void main(String[] args) {
        Company c = new Company("TTT", "广州市", "000000-0000-000000",
            2000, 10, 10);
        c.info();
    }
}
```

运行这个程序，显示以下信息。

```
公司名称：TTT
公司地址：广州市
公司纳税人号码：000000-0000-000000
公司成立日期：2000-10-10
```

由于 FoundDate 类是 Company 类的成员内部类，FoundDate 类是依附于 Company 对象而存在的，因此，不能在其他外部类直接创建 FoundDate 类的对象。例如，不能在 Main 类中直接创建 FoundDate 类的对象。

4.4.2 静态内部类

静态内部类与成员内部类有些类似，只是在静态内部类的定义中使用了 static 关键字对内部类的定义进行了限定，因此，静态内部类类似于静态属性：静态内部类是依附于外部类而不是外部类的对象而存在的,因此,可以在其他外部类中直接创建静态内部类的对象。

下面举一个例子说明静态内部类的使用。定义一个用于 Person 类的 Java 类。这个类包含如下属性：name 表示姓名，idd 表示身份证号码，revenue 表示收入情况，address 表示联系地址；包括如下方法：构造函数、信息显示方法。同时，为了能够更为精细地表示 idd 身份证号码、revenue 年收入情况两个属性的值，定义了两个静态内部类。完整的 Person 类代码如下（源代码为 04-20.java）：

```java
package org.example;
import java.awt.desktop.SystemEventListener;
public class Person {
    String name;
    IDD idd;
    Revenue revenue;
    String address;
    public Person(String name, IDD idd, Revenue revenue, String address) {
        this.name = name;
        this.idd = idd;
        this.revenue = revenue;
        this.address = address;
    }
    public void info() {
        System.out.println("姓名：" + name);
        System.out.println("IDD：" + idd.toString());
        System.out.println("Revenue: " + revenue);
        System.out.println("联系地址：" + address);
    }
    public static class IDD {
        String province;
        String birth;
        String area;
        public IDD(String province, String birth, String area) {
            this.province = province;
            this.birth = birth;
            this.area = area;
        }
```

```
        public String toString() {
            return province+birth+area;
        }
    }
    public static class Revenue {
        float gaining;      // 年收入
        float tax;          // 纳税额
        float education;    // 教育支出
        public Revenue(float gaining, float tax, float education) {
            this.gaining = gaining;
            this.tax = tax;
            this.education = education;
        }
        public String toString() {
            return "年收入: " + gaining + ", 纳税额: " + tax + ", 教育支出: " + education;
        }
    }
}
```

在 Person 类的定义中，包含了两个静态内部类：IDD 类和 Revenue 类，分别表示身份证信息以及年收入和支出情况。由于它们都是静态内部类，并且都使用了 public 访问限定符，可以在其他类中创建这两个静态内部类的对象。创建静态内部类对象的一般形式如下：

外部类名.静态内部类 变量 = new 外部类名.静态内部类(构造函数参数);

现在，修改 Main 类，创建 IDD 类和 Revenue 类的对象，并进一步创建 Person 类的对象，修改后的 Main 类代码如下（源代码为 04-21.java）：

```java
package org.example;
public class Main {
    public static void main(String[] args) {
        Person.IDD idd = new Person.IDD("900101", "20201010", "2201");
        Person.Revenue revenue = new Person.Revenue(153201.5f,
        120000.0f, 30000.0F);
        Person p = new Person("张三", idd, revenue, "广州市");
        p.info();
    }
}
```

在 Main 类的 main( ) 方法中，先创建 IDD 和 Revenue 静态内部类的对象，然后创建 Person 类的对象，并显示 Person 对象的信息。运行这个程序，显示以下信息。

姓名:张三

```
IDD：9001012020101022201
Revenue：年收入：153201.5，纳税额：120000.0，教育支出：30000.0
联系地址：广州市
```

4.4.3 使用内部类应用实践

使用成员内部类的应用实践是：如果一个类仅在一个特定的外部类中使用，并且这个内部类不需要对其他外部类可见，则将这个类定义为成员内部类。

使用静态内部类的应用实践是：如果一个外部类（如 Person 类）需要使用到几个关联的内部类（如 IDD 类和 Revenue 类），并且这几个内部类除了对这个特定的外部类可见外，还可以对其他外部类可见，则将这几个内部类定义为静态内部类。

4.5 案例：使用 Java 类描述一元二次方程

除了可以使用 Java 类对实体对象进行描述外，如在前面的例子中使用 Java 类描述 Vehicle、Student 类等，还可以使用 Java 类对虚拟的抽象对象进行描述。本节通过一个例子介绍如何使用 Java 类描述一些抽象对象，即描述一元二次方程。

4.5.1 案例任务

编写一个 Java 程序，该程序可定义一个能够描述任意一元二次方程，并包含能够计算一元二次方程根的方法。再编写一个测试类，要求首先输入一元二次方程的三个系数，然后计算并输出这个一元二次方程的根。

4.5.2 任务分析

数学上，一元二次方程的一般形式为 $ax^2+bx+c=0$。从这种表达形式可知，只要给出 a、b、c 三个系数的值，则可以唯一确定一个一元二次方程。因此，可以定义一个名为 Quadratic 的类，这个类中包含 a、b、c 三个属性，同时定义用于计算一元二次方程根的方法即可。

4.5.3 任务实施

在 ch04-01 工程的 org.example 包下，新建一个名为 Quadratic 的 Java 类，然后打开 Quadratic 类，输入以下代码（源代码为 04-22.java）。

```
package org.example;
```

```java
public class Quadratic {
    float a, b, c;
    public Quadratic(float a, float b, float c) {
        this.a = a;
        this.b = b;
        this.c = c;
    }
    public void getRoots() {
        if (a == 0) {
            if (b == 0) {
                if (c == 0) {
                    System.out.println("任何实数都是所给一元二次方程的根");
                    return;
                }
                else {
                    System.out.println("所给系数不能构成一元二次方程");
                    return;
                }
            }
            System.out.println("所给系数构成的一元二次方程只有一个根:" + (-c/b));
            return;
        }
        float delta = b*b - 4*a*c;
        if (delta > 0) {
            System.out.println("所给一元二次方程有两个不同实根：");
            System.out.println((-b + Math.sqrt(delta)/(2*a)));
            System.out.println((-b - Math.sqrt(delta)/(2*a)));
        }
        else if (delta == 0) {
            System.out.println("所给一元二次方程有两个相同实根：");
            System.out.println((-b)/(2*a));
        }
        else {
            System.out.println("所给一元二次方程没有实数根");
        }
    }
    public String toString() {
        return "一元二次方程:" + a + "x^2 + " + b + "x + " + c;
    }
}
```

注意：代码中用到了 JDK 中定义的求实数平方根的函数 Math.sqrt(实数参数)。在后面章节中会介绍这个函数的使用。

为了测试 Quadratic 类代码的正确性，修改 Main 类代码，先输入三个系数，然后创建 Quadratic 类的对象，并计算一元二次方程的根。修改后的 Main 代码如下（源代码为 04-23.java）：

```java
package org.example;
import java.util.Scanner;
public class Main {
    public static void main(String[] args) {
        float a, b, c;
        Scanner sc = new Scanner(System.in);
        System.out.print("请输入系数a:");
        a = sc.nextFloat();
        System.out.print("请输入系数b:");
        b = sc.nextFloat();
        System.out.print("请输入系数c:");
        c = sc.nextFloat();
        Quadratic q = new Quadratic(a, b, c);
        System.out.println(q.toString());
        q.getRoots();
    }
}
```

运行这个程序，显示以下结果。

```
请输入系数a:1
请输入系数b:10
请输入系数c:5
一元二次方程: 1.0x^2 + 10.0x + 5.0
所给一元二次方程有两个不同实根：
-5.52786404500042
-14.47213595499958
```

4.6 练习：计算三角形的面积和周长

编写一个 Java 程序，该程序首先定义一个能够表示任意三角形的类。再编写一个测试类，要求首先输入三角形的三条边长，然后构建三角形对象，最后计算并输出三角形的面积和周长。

第 5 章 继承和多态

类是对同类事物对象共有性质的封装：首先将同类事物封装为类，然后可以通过 new 关键字创建类的多个对象。在面向对象程序设计中，封装是一个重要概念。与封装的重要性类似，继承与多态也是面向对象程序设计中的重要概念。本章对继承和多态进行介绍。

5.1 类 的 继 承

类之间可能存在继承关系。例如，用于描述学生的 Student 类、用于描述老师的 Teacher 类和用于描述工人的 Worker 类都是一个更为宽泛的 Person 类的特殊情况，因为，无论是学生、老师还是工人，他们首先都是人。对于这种现象，是否可以采用一种更为有效的方式来定义 Person 类、Student 类、Teacher 类和 Worker 类及其之间的关系呢？也就是说，是否可以在 Person 类中定义 Student 类、Teacher 类和 Worker 类所共有的属性和方法，然后，通过一种技术手段，使 Student 类、Teacher 类和 Worker 类可以从 Person 类中继承这些共有的属性和方法，从而简化类的设计呢？在 Java 中确实存在这种机制，这种机制就是类的继承。

5.1.1 继承的概念

图 5-1 描述了 Person 类与 Student 类、Teacher 类、Worker 类之间的继续关系。

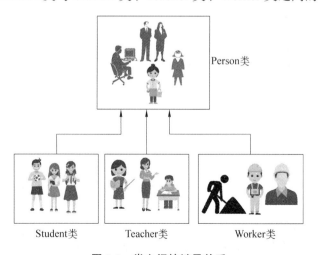

图 5-1 类之间的继承关系

对于 Person 类而言，都具有这些基本属性：姓名、身份证号码、联系地址、出生日期等，同时还具有以下基本方法：显示人的基本信息。对于 Student 类而言，不仅具有 Person 类的基本属性和方法外，而且有自己独特的属性，包括学号、所在学校名称等，还具有自己的用于显示信息的方法等。Teacher 类和 Worker 类也是类似的。

在 Java 中，可以使用继承这种技术简化类的设计，针对这里的 Person 类与 Student 类、Teacher 类、Worker 类的设计，可以先定义 Person 类，然后使用继承技术定义 Student 类、Teacher 类、Worker 类，从而提高类的编码效率。在面向对象概念中，Person 类称为父类，Student 类、Teacher 类、Worker 类称为 Person 类的子类。

5.1.2 定义类的继承关系

为了定义 Person 父类及其子类 Student 类、Teacher 类、Worker 类，先在 IDEA 中新建一个名为 ch05-01 的工程，如图 5-2 所示。

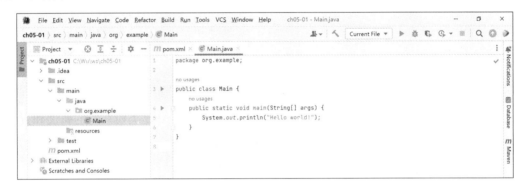

图 5-2 新建的 ch05-01 工程

在 IDEA 创建包和创建子类

如前所述，一个规划良好的 Java 工程应该将 Java 代码分门别类地放置在不同的程序包中。为此，在 ch05-01 工程下，新建一个名为 com.ttt.human 的程序包，如图 5-3 所示。在图 5-3 中，选择 java → New → Package 命令，在弹出的窗口中输入 com.ttt.human，从而新建了 com.ttt.human 的程序包。

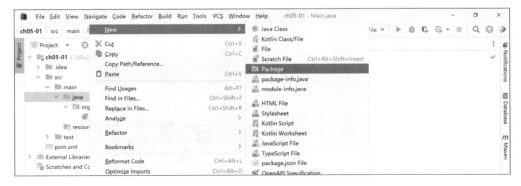

图 5-3 新建 com.ttt.human 程序包

在新建 com.ttt.human 程序包后，在这个包中新建名为 Person 的 Java 类，在 Person 类中输入以下 Person 类的代码（源代码为 05-01.java）。

```java
package com.ttt.human;
public class Person {
    String name;
    String id;   // 身份证号码
    String address;
    public Person() {
        this.name = "Unknown";
        this.id = "Unknown";
        this.address = "Unknown";
    }
    public Person(String name, String id, String address) {
        this.name = name;
        this.id = id;
        this.address = address;
    }
    public void setName(String name) {
        this.name = name;
    }
    public void setId(String id) {
        this.id = id;
    }
    public void setAddress(String address) {
        this.address = address;
    }
    public void display() {
        System.out.println("姓名:" + name);
        System.out.println("身份证号码:" + id);
        System.out.println("联系地址:" + address);
    }
    public String toString() {
        return "姓名:" + name + ", 身份证号码:" + id + ", 联系地址:" + address;
    }
}
```

完成了父类 Person 类的定义后，可以基于 Person 类定义 Student 子类、Teacher 子类和 Worker 子类。在 Java 中，基于已有的父类定义子类的一般形式如下：

```java
public class 子类 extends 父类 {
    //子类的属性
    //子类的方法
}
```

先定义 Student 类。Student 类除了包含 Person 类的属性和方法外，还具有以下这些特殊的属性：学号、所在学校名称。同时，对于 Student 类中的 display( ) 方法，除了要显示姓名、联系地址、身份证号码以外，还要显示学号、所在学校名称等学生的基本信息，因此，需要重写 Student 类的 display( ) 方法。在 ch05-01 工程的 com.ttt.human 包下，新建名为 Student 的 Java 类。Student 类的代码如下（源代码为 05-02.java）：

```java
package com.ttt.human;
public class Student extends Person {
    String sno;
    String school;
    public Student(String name, String id, String address, String sno,
    String school) {
        this.name = name;
        this.id = id;
        this.address = address;
        this.sno = sno;
        this.school = school;
    }
    @Override
    public void display() {
        System.out.println("姓名:" + name);
        System.out.println("身份证号码:" + id);
        System.out.println("联系地址:" + address);
        System.out.println("学号:" + sno);
        System.out.println("所在学校:" + school);
    }
    @Override
    public String toString() {
        return "姓名:" + name + ",身份证号码:" + id + ",联系地址:" +
address +
            ",学号:" + sno + ",所在学校:" + school;
    }
}
```

在定义 Student 类时，使用了以下语句。

```java
public class Student extends Person {
```

表示 Student 类是 Person 类的子类。然后，定义了 Student 类特有的属性和方法。这里需要强调一下，Java 只支持单继承，也就是说，一个子类只能有一个父类。具体来说就是，在 extends 关键字之后只能有一个父类名。

在 Student 类的代码中，由于 display( ) 方法和 toString( ) 方法已经在其父类 Person 类中定义过，在 Student 类中，根据需要对这两个方法进行了重写，为了告诉 Java 这一事实，使用了 Java 的注解机制：通过 @Override 注解告诉 Java，在 Student 类中对父类的同名方法进行了重写。

通过 Student 类的代码可以发现，虽然 Student 类是 Person 类的子类，然而，在 Student 类的定义中并没有充分利用继承技术所带来的代码编写效率。具体来说，观察一下 Student 类的构造函数。

```java
public Student(String name, String id, String address, String sno, String school) {
    this.name = name;
    this.id = id;
    this.address = address;
    this.sno = sno;
    this.school = school;
}
```

注意其中的：

```java
this.name = name;
this.id = id;
this.address = address;
```

显然，这些语句完成对属于 Person 类属性的初始化赋值工作。观察一下 Person 类的代码可以注意到，在 Person 类中存在一个 Person 的构造函数完成完全相同的工作。

```java
public Person(String name, String id, String address) {
    this.name = name;
    this.id = id;
    this.address = address;
}
```

类似的情况还出现在 Student 类对 Person 类的重写方法中。例如，在 Student 类的 display() 方法中，部分代码与父类中的代码完全一致。

```java
@Override
public void display() {
    System.out.println("姓名:" + name);      // 以下3行代码与父类的display()
                                              //     完全一致
    System.out.println("身份证号码:" + id);
    System.out.println("联系地址:" + address);
    System.out.println("学号:" + sno);
    System.out.println("所在学校:" + school);
}
```

在 Java 中，利用继承，可以在使子类继承父类属性和方法的同时，能够通过使用 super 关键字提高编码效率。

5.1.3　super 关键字及方法重写

super 关键字的主要作用：在子类的构造函数中使用 super 关键字时，super 指称父类的构造函数；在子类的方法中使用 "super.父类方法" 的形式，可以在子类中调用父类的指定方法。现在使用 super 关键字重写 Student 子类。其完整代码如下（源代码为 05-03.java）：

```java
package com.ttt.human;
public class Student extends Person {
    String sno;
    String school;
    public Student(String name, String id, String address, String sno,
    String school) {
        super(name, id, address);
        this.sno = sno;
        this.school = school;
    }
    @Override
    public void display() {
        super.display();
        System.out.println("学号：" + sno);
        System.out.println("所在学校：" + school);
    }
    @Override
    public String toString() {
        return super.toString() +
            ", 学号：" + sno + ", 所在学校：" + school;
    }
}
```

在这个新的 Student 类定义中使用了 super 关键字，简化了 Student 类的编码：使用 super 关键字调用父类的构造函数，从而简化了 Student 类的构造函数的代码；在子类方法的重写中，使用 "super.父类方法" 的形式调用父类的方法，简化了子类重写父类方法的代码，使代码更为简洁和清晰。

现在修改 Main 类的代码，在 main() 函数中创建 Student 类的对象，并显示 Student 对象的信息。然后，调用 Student 类从 Person 类继承的 setName() 方法，修改 Student 对象的 name 属性值，再一次调用 display() 方法显示 Student 对象的信息。Main 类的代码如下（源代码为 05-04.java）：

```java
package org.example;
import com.ttt.human.Student;
```

```
public class Main {
    public static void main(String[] args) {
        Student student = new Student("Wanger", "91010119991010××××",
        "广州市",
                "202200001234", "英才中学");
        student.display();
        student.setName(" 王二 ");
        student.display();
    }
}
```

在 Main 类的代码中，由于使用到了 Student 类，并且由于 Student 类与 Main 类不在同一个包中，所以，需要使用以下 import 语句导入 Student 类。

import com.ttt.human.Student;

然后，在 main( ) 函数中使用语句。

```
Student student = new Student("Wanger", "91010119991010××××", "广州市",
        "202200001234", "英才中学");
student.display();
```

创建了 Student 类的对象，并且使用 student 变量引用这个对象，再调用 student 对象的 display( ) 方法显示 student 对象的信息。之后，使用语句：

student.setName(" 王二 ");

修改了 student 对象的属性值，并再次调用 student 对象的 display( ) 方法显示修改后的 student 对象的信息。注意到，Student 类中并没有定义 setName( ) 方法，但是在这里却使用了 setName( ) 方法修改 name 属性的值，这是因为 Student 类从 Person 类中继承了 setName( ) 方法。这就是继承的本质。运行这个程序，显示以下运行结果。

```
姓名: Wanger
身份证号码: 91010119991010××××
联系地址: 广州市
学号: 202200001234
所在学校: 英才中学
姓名: 王二
身份证号码: 91010119991010××××
联系地址: 广州市
学号: 202200001234
所在学校: 英才中学
```

5.1.4 练习：完成 Teacher 子类和 Worker 子类的代码编写

参考 Student 类的代码，以 Person 类为父类，完成 Teacher 子类和 Worker 子类的代码编写并编写一个 Main 类以测试 Teacher 子类和 Worker 子类代码的正确性。要求使用继承和 super 关键字，使编写的代码更为简洁和清晰。

5.2 访问限定符

如前所述，通过类将同类事物的属性和方法封装起来，并使该类的对象都具有类似的属性和方法。在类的封装技术中，还有一个非常重要的概念，那就是访问限定符：用于限定哪些类可以访问对象的属性和方法。

5.2.1 访问限定符及其可访问性

Java 提供了四个访问限定符：公开（public）、保护（protected）、默认（default）、私有（private）。这些访问限定符既可以对属性也可以对方法的可访问性进行控制。这四个访问限定符对属性和方法的访问控制如表 5-1 所示。

表 5-1 访问限定符对属性和方法的访问控制

名称	修饰符	同一类	同一包	子类	不同包
公开	public	可访问	可访问	可访问	可访问
保护	protected	可访问	可访问	可访问	×
默认	default	可访问	可访问	×	×
私有	private	可访问	×	×	×

1. public 访问限定符

被 public 访问限定符修饰的成员可以被任何类访问到。public 访问限定符可以修饰类、构造方法、属性成员、方法成员。

2. protected 访问限定符

被 protected 访问限定符修饰的成员既可以被同一个类的成员访问，也可以被同一包中的其他类访问，还可以被子类访问。protected 访问限定符可以修饰属性成员、构造方法、方法成员。

3. default 访问限定符

被 default 访问限定符修饰的成员既可被同一个类的成员访问，也可以被同一包中其他类访问。default 访问限定符可以修饰类、构造方法、属性成员、方法成员。

4. private 访问限定符

被 private 访问限定符修饰的成员只能在当前类中被访问到。private 访问限定符可以修饰构造方法、属性成员、方法成员。

5.2.2 访问限定符使用举例

通过一个例子来介绍访问限定符的概念和使用。在 ch05-01 工程下新建一个名为 com.ttt.modifier 的包,在这个包下新建两个 Java 类:One 类和 Two 类,同时,在 org.example 包下新建 One 类的子类 SubOne。在这个例子中,对不同访问限定符的使用进行举例说明。

访问限定符使用举例

5.3 抽象类和多态

到目前为止,本章定义的各种 Java 类,如 Student 类、Person 类、One 类等都是可实例化的类,也就是基于这些类可以使用 new 关键字创建类的对象。在 Java 中还可以定义另一种类型的类,称为抽象类。所谓抽象类,就是不能使用 new 关键字创建实例对象的类。抽象类一般用于在类的继承中作为父类而存在,并且与多态有密切关系。

5.3.1 抽象类和使用 final 关键字修饰属性

观察如图 5-4 所示的一组图形形状,这组图形中包括三角形、矩形和圆形。虽然它们的外观各异,但是它们都有一些共同的属性和方法。例如,任何图形形状都有名字,可以计算任何图形形状的周长和面积等。

图 5-4 一组图形形状

考虑到图形形状都有一些共同的属性和方法,为了提高编码效率和代码重用性,可以先定义一个 Shape 类来表示图形形状的共有属性和方法。为此,在 ch05-01 工程下新建一个名为 com.ttt.shapes 的包,在这个包下新建 Shape 类。Shape 类的代码如下(源代码为 05-05.java):

```
package com.ttt.shapes;
public class Shape {
    public static final float PI = 3.14f;
    private String name;
    public Shape(String name) {
```

```
        this.name = name;
    }
    public String getName() {
        return name;
    }
    public void setName(String name) {
        this.name = name;
    }
    public float getCircumference() {
        // 因为不知道是何种图形，没法计算周长
        return 0.0f;
    }
    public float getArea() {
        // 因为不知道是何种图形，没法计算面积
        return 0.0f;
    }
}
```

在 Shape 类中定义了一个名为 PI 的静态 static 的 public 属性，同时使用了 final 关键字对这个 PI 属性进行了修饰。

```
public static final float PI = 3.14f;
```

对这句代码的准确完整的理解是：public 访问限定符表示 PI 这个属性在任何其他类中是可以被访问的；static 修饰符则表示 PI 这个属性属于 Shape 这个类，因此可以使用 Shape.PI 来访问这个属性的值；final 修饰符则表示 PI 这个属性值一旦被初始化赋值，不能再以任何方式改变。

注意 Shape 类中用于计算图形周长和面积的方法。由于不知道是何种图形，因此，目前只能使用这种"粗劣"的直接返回 0 的方式定义计算周长和面积的方法。

```
public float getCircumference() {
    // 因为不知道是何种图形，没法计算周长
    return 0.0f;
}
public float getArea() {
    // 因为不知道是何种图形，没法计算面积
    return 0.0f;
}
```

Java 语言的设计者已经解决了这个问题：将 Shape 类定义为抽象类。修改后的 Shape 类代码如下（源代码为 05-06.java）：

```
package com.ttt.shapes;
```

```
public abstract class Shape {
    public static final float PI = 3.14f;
    private String name;
    public Shape(String name) {
        this.name = name;
    }
    public String getName() {
        return name;
    }
    public void setName(String name) {
        this.name = name;
    }
    // 因为不知道是何种图形，没法计算周长，所以将这个方法定义为抽象的
    abstract public float getCircumference();
    // 因为不知道是何种图形，没法计算面积，所以将这个方法定义为抽象的
    abstract public float getArea();
}
```

因为 Shape 类包含抽象的方法：

```
abstract public float getCircumference();
```

和

```
abstract public float getArea();
```

所以 Shape 类也必须声明为抽象的。

```
public abstract class Shape {
```

因为 Shape 类是抽象类，所以不能使用 new 关键字创建 Shape 类的对象。原因很简单：如果能够创建 Shape 类的对象，试想一下，如果调用所创建对象的 getCircumference( ) 方法或者 getArea( ) 方法会发生什么情况呢？结果是不可预知的。因为，这两个方法根本就没有具体功能代码。

在面向对象设计理念中，抽象类都是用在继承中作为父类而存在的。因此，基于 Shape 类，现在可以定义具体的三角形类、圆形类和矩形类。

先定义三角形类 Triangle 类。因为三条边长可以唯一确定一个三角形，所以，三角形类继承 Shape 类，增加三个边长属性，并重写计算周长和面积的方法即可。Triangle 类的代码如下（源代码为 05-07.java）：

```
package com.ttt.shapes;
public class Triangle extends Shape {
```

```java
    private float a, b, c;
    public Triangle(String name, float a, float b, float c) {
        super("三角形");
        this.a = a;
        this.b = b;
        this.c = c;
    }
    @Override
    public float getCircumference() {
        return a+b+c;
    }
    @Override
    public float getArea() {
        float p = (a+b+c)/2;
        return (float) Math.sqrt(p*(p-a)*(p-b)*(p-c));
    }
}
```

在 Triangle 类中，由于已经实现了父类的抽象方法 getCircumference( ) 和 getArea( )，因此，Triangle 类不再是抽象的。在 Triangle 类的 getArea( ) 方法中，使用了 Java 的强制转换操作。

```java
@Override
public float getArea() {
    float p = (a+b+c)/2;
    return (float) Math.sqrt(p*(p-a)*(p-b)*(p-c));
}
```

这里需要做强制转换操作的原因是：因为 getArea( ) 方法的返回值必须是 float 类型的，而 Math.sqrt( ) 方法的返回值是 double 类型的，所以，需要使用语句：

```
(float) Math.sqrt(p*(p-a)*(p-b)*(p-c))
```

将 double 类型强制转换为 float 类型，这样可以避免 IDEA 告警。

类似地，可以定义圆 Circle 类和矩阵 Rectangle 类。由于这两个类的定义都比较简单，不用对代码做过多解释。Circle 类的代码如下（源代码为 05-08.java）：

```java
package com.ttt.shapes;
public class Circle extends Shape {
    private float r;
    public Circle(float r) {
        super("圆形");
        this.r = r;
```

```
    }
    @Override
    public float getCircumference() {
        return 2*PI*r;
    }
    @Override
    public float getArea() {
        return PI*r*r;
    }
}
```

Rectangle 类的代码如下（源代码为 05-09.java）：

```
package com.ttt.shapes;
public class Rectangle extends Shape {
    private float a, b;
    public Rectangle(float a, float b) {
        super(" 矩形 ");
        this.a = a;
        this.b = b;
    }
    @Override
    public float getCircumference() {
        return 2*(a+b);
    }
    @Override
    public float getArea() {
        return a*b;
    }
}
```

上面已经定义了四个类：Shape 类、Triangle 类、Circle 类和 Rectangle 类。由于 Shape 是抽象类，不能创建这个类的对象，但是可以定义 Shape 类型的变量，并可以使 Shape 类型的变量引用 Triangle 类、Circle 类和 Rectangle 类的对象。这种现象称为多态。

5.3.2 多态

在 Java 中，使用父类变量引用子类对象的现象称为多态，也就是说，父类变量可以引用子类的任何对象。例如，可以定义 Shape 的变量 shape，然后，使 shape 变量引用 Triangle 对象。为了例示这种现象，修改 Main 类，在 main( ) 方法中定义 Shape 类的变量，并使 Shape 类的变量引用 Triangle 对象。修改后的 Main 类代码如下（源代码为 05-10.java）：

```
package org.example;
import com.ttt.shapes.Shape;
import com.ttt.shapes.Triangle;
public class Main {
    public static void main(String[] args) {
        Shape shape = new Triangle(10.0f, 15.0f, 10.0f);
        System.out.println("图形名称为: " + shape.getName());
        System.out.println("三角形的周长: " + shape.getCircumference());
        System.out.println("三角形的面积: " + shape.getArea());
    }
}
```

运行这个程序，显示以下结果。

```
图形名称为：三角形
三角形的周长：35.0
三角形的面积：49.607838
```

5.3.3 使用 instanceof 关键字检查对象类型

由于多态现象，有时需要判定父类变量所引用的子类对象所属的类。例如，Shape 类的变量 shape 可以引用任何 Shape 子类对象，如何才能知道 shape 变量当前所引用的对象属于哪个类呢？Java 提供了 instanceof 关键字用于解决这个问题。例如，下面这个 Main 类代码将根据 shape 变量当前所引用的对象类型，将 shape 引用的对象强制转换为相应类的变量并显示类型信息（源代码为 05-11.java）。

```
package org.example;
import com.ttt.shapes.Circle;
import com.ttt.shapes.Rectangle;
import com.ttt.shapes.Shape;
import com.ttt.shapes.Triangle;
public class Main {
    public static void main(String[] args) {
        Shape shape = new Triangle(10.0f, 15.0f, 10.0f);
        if (shape instanceof Circle) {
            Circle c = (Circle) shape;
            System.out.println("shape 引用的对象是圆形类");
        }
        else if (shape instanceof Rectangle) {
            Rectangle r = (Rectangle) shape;
            System.out.println("shape 引用的对象是矩形类");
        }
```

```
        else if (shape instanceof Triangle) {
            Triangle t = (Triangle) shape;
            System.out.println("shape 引用的对象是三角形类");
        }
    }
}
```

运行这个程序,将显示以下信息。

shape 引用的对象是三角形类

更为简洁地,上面这个 Main 类代码可以简化为以下形式(源代码为 05-12.java)。

```
package org.example;
import com.ttt.shapes.Circle;
import com.ttt.shapes.Rectangle;
import com.ttt.shapes.Shape;
import com.ttt.shapes.Triangle;
public class Main {
    public static void main(String[] args) {
        Shape shape = new Triangle(10.0f, 15.0f, 10.0f);
        if (shape instanceof Circle c) {
            System.out.println("shape 引用的对象是圆形类");
        }
        else if (shape instanceof Rectangle r) {
            System.out.println("shape 引用的对象是矩形类");
        }
        else if (shape instanceof Triangle t) {
            System.out.println("shape 引用的对象是三角形类");
        }
    }
}
```

如加黑代码所示:在使用 instanceof 关键字判定 shape 变量的类型后,直接将 shape 引用的对象强制转换为目标类的变量引用。

5.3.4 对象数组

第 3 章介绍了数组定义和使用。对于基本数据类型,如 int 类型、float 类型等,可以使用以下方式定义并初始化一个数组。

```
float[] scores;
scores = new float[20];
```

然后通过下标即可访问数组的元素。例如，采用下面的代码可将数组第 0 个元素的值设置为 90.5f。

```
scores[0] = 90.5f;
```

对于 Java 类，也可以定义数组。例如，下面的代码定义了 Shape 类的数组，并且这个数组有 10 个元素。

```
Shape[] shape;
shape = new Shape[10];
```

由于目前 Shape 数组中的每个元素只是 Shape 类型的一个变量。注意，每个元素仅是一个变量而已：数组元素的变量都还没有引用到具体的对象，所以此时还不能使用数组元素；否则，Java 运行时会出现错误。

为了能够正常使用数组元素，必须使每个数组元素引用到合法的对象，为此，对于每个数组元素，还需要使用 new 关键字。例如，可以使用下面的语句使第 0 个数组元素引用到合法对象。

```
shape[0] = new Triangle(10.0f, 13.0f, 10.0f);
```

此时，shape[0] 已经引用到了合法的三角形对象。因此，可以通过 shape[0] 来使用这个三角形对象。例如，计算这个三角形的周长或者面积。

```
shape[0].getCircumference();
shape[0].getArea();
```

下面通过一个例子介绍如何定义对象数组，并使用 instanceof 关键字来判定数组中每个对象所属的类。该程序先生成一个随机数，并根据随机数不同使 Shape 数组的每个元素引用不同类型的图形对象，然后在另一个循环中使用 instanceof 关键字判定数组元素所引用对象的类型并显示信息。Main 类代码如下（源代码为 05-13.java）：

```
package org.example;
import com.ttt.shapes.Circle;
import com.ttt.shapes.Rectangle;
import com.ttt.shapes.Shape;
import com.ttt.shapes.Triangle;
import java.util.Random;
public class Main {
    public static void main(String[] args) {
        Shape[] shapes;
        shapes = new Shape[10];
        Random random = new Random();
```

```
        for(int i=0; i<10; i++) {
            int which = random.nextInt(3);
            shapes[i] = switch(which){
                case 0 -> new Triangle(random.nextInt(), random.
                nextInt(), random.nextInt());
                case 1 -> new Circle(random.nextInt());
                case 2 -> new Rectangle(random.nextInt(), random.
                nextInt());
                default -> null;
            };
        }
        for(int i=0; i<shapes.length; i++) {
            if (shapes[i] instanceof Triangle t)
                System.out.println(" 数组中的这个元素对象是三角形 ");
            else if (shapes[i] instanceof Circle c)
                System.out.println(" 数组中的这个元素对象是圆形 ");
            else if (shapes[i] instanceof Rectangle r)
                System.out.println(" 数组中的这个元素对象是矩形 ");
        }
        System.out.println("Over");
    }
}
```

在 Main 代码中，语句：

```
Random random = new Random();
```

用于创建一个可以生成随机数的对象。Random 类是 JDK 定义的用于生成随机数的类。在创建类随机数对象后，使用语句：

```
int which = random.nextInt(3);
```

生成一个在 0~2 内的整数随机数，并根据这个随机数的值使用 switch 表达式使 shapes 数组的每个元素引用不同的图形形状。最后，通过 for 循环判定每个数组元素所引用的对象类型并显示结果。

```
for(int i=0; i<shapes.length; i++) {
    if (shapes[i] instanceof Triangle t)
        System.out.println(" 数组中的这个元素对象是三角形 ");
    else if (shapes[i] instanceof Circle c)
        System.out.println(" 数组中的这个元素对象是圆形 ");
    else if (shapes[i] instanceof Rectangle r)
        System.out.println(" 数组中的这个元素对象是矩形 ");
}
```

注意 for 循环中的用于获取数组元素个数的语句,即 **shapes.length**。一旦创建了一个数组,就可使用"**数组名 .length**"获取数组元素的个数。运行这个程序,显示以下结果。

```
数组中的这个元素对象是三角形
数组中的这个元素对象是三角形
数组中的这个元素对象是圆形
数组中的这个元素对象是三角形
数组中的这个元素对象是三角形
数组中的这个元素对象是圆形
数组中的这个元素对象是矩形
数组中的这个元素对象是三角形
数组中的这个元素对象是三角形
数组中的这个元素对象是圆形
Over
```

注意:由于生成的随机数不同,每次运行该程序的结果可能会有所差异。

5.4 使用 final、record 和 sealed 关键字修饰类

在 Java 编程实践中,可能需要对某些类的定义做一些限制。例如,限制某个类不能作为父类使用;限制一旦创建了某个类的对象,则不能修改对象的属性值等。为了满足这些要求,Java 提供了必要的关键字。

5.4.1 使用 final 关键字修饰类

final 关键字可以用来修饰类、属性和方法。顾名思义,final 就是"最后"的意思。因此,使用 final 关键字修饰的属性一旦被初始化,则在后续的任何时候都不可以再次修改该属性的值。可以在两个地方对用 final 修饰的属性进行初始化:一是其定义处,也就是说在 final 属性定义时直接给其赋值;二是在构造函数中,二者只能选其一。如果使用 final 关键字修饰了某个类的方法,则意味着在将该类作为父类使用时,在子类中不能重写在父类中被 final 关键字修饰的方法。使用 final 关键字修饰的类不能被继承。

5.4.2 使用 record 关键字定义 Java 类

record 关键字用于创建不可变的数据类:使用 record 关键字可以快速简洁地定义属性不可变更的 Java 类。当使用 record 关键字声明一个类时,Java 将为该类自动创建这些方法:构造方法,用于创建对象;hashCode( ) 方法,用于为该类的每个对象生成一个 hash 码;euqals( ) 方法,用于比较该类的两个对象是否相等;toString( ) 方法,用于将该类的对象转换为一个字符串。同时,使用 record 关键字修饰的类被 final 关键字修饰,从而不能被继承。

下面举一个例子说明 record 关键字的使用。在 Main 类中，创建一个用 record 关键字修饰的类 Person，然后，在 Main 类的 main( ) 方法中创建 Person 类的对象。Main 类的代码如下（源代码为 05-14.java）：

```
package org.example;
public class Main {
    private record Person(String name, int age){}
    public static void main(String[] args) {
        Person p1 = new Person("张三", 20);
        System.out.println(p1.age);
        System.out.println(p1.name);
        System.out.println(p1.hashCode());
        System.out.println(p1.toString());
        Person p2 = new Person("张三", 20);
        System.out.println(p2.age);
        System.out.println(p2.name);
        System.out.println(p2.hashCode());
        System.out.println(p2.toString());
        Person p3 = p1;
        System.out.println(p1 == p2);      // 显示 false
        System.out.println(p1 == p3);      // 显示 true
    }
}
```

在 Main 类的代码中，通过语句：

```
private record Person(String name, int age){}
```

定义了一个 private 的 record 类 Person，该类包含 name 和 age 两个属性。然后在 main( ) 函数中使用语句：

```
Person p1 = new Person("张三", 20);
System.out.println(p1.age);
System.out.println(p1.name);
System.out.println(p1.hashCode());
System.out.println(p1.toString());
```

创建了 Person 类的对象并显示相应信息。然后，使用语句：

```
Person p2 = new Person("张三", 20);
System.out.println(p2.age);
System.out.println(p2.name);
System.out.println(p2.hashCode());
System.out.println(p2.toString());
```

创建了 Person 类的对象并显示相应信息。这从下面的代码可以得出结论。

```
Person p3 = p1;
System.out.println(p1 == p2);    // 显示 false
System.out.println(p1 == p3);    // 显示 true
```

运行这个程序，将显示以下结果。

```
20
张三
24021579
Person[name=张三, age=20]
20
张三
24021579
Person[name=张三, age=20]
false
true
```

5.4.3 使用 sealed 关键字修饰类

sealed，顾名思义，就是"密封"的意思。在面向对象语言中，可以通过继承实现类的能力复用、扩展与增强。但有时不希望一个类被其他类无限制地继承，因此，需要对继承关系有一些限制手段。而密封类的作用就是限制类的继承。密封类可以控制哪些类可以对超类进行继承：使用 sealed 关键字和 permits 关键字即可实现这种限制。下面举例说明 sealed 关键字的使用。

使用 sealed 关键字修饰类

5.5 案例：定义 Java 程序类应用实践

5.5.1 案例任务

Java 提供了灵活地定义 Java 类的手段。从工程实践角度看，Java 类大致包括以下几种类型：用于封装对象实体的 Java 类（persistent object, PO）、用于程序功能部件之间进行数据交互的 Java 类（value object, VO）、用于完成某些特定功能的 Java 功能类（function object, FO）。

5.5.2 任务分析

PO 类是 Java 中非常简单但非常常用的一种类，主要用于封装实体对象，如用于封装学生实体对象、书籍实体对象等。这类对象经常用于与数据库的交互。PO 类，有时也称为POJO类，也就是Plain Old Java Object。这些对象都有明确的属性，并且可以修改及读取实体对象的属性值，因此，这类对象除了有构造函数外，都有setter( )、getter( )方法和一些其他的辅助方法。例如，toString( )方法用于将对象属性字符串化，hashCode( )方法用于产生对象的hash码等。

VO类是一种一旦创建了这个类的对象就不能再修改对象属性的特殊Java类，也是常用的Java类之一。例如，在使用Java编写基于Web的网络应用时，在Java程序中创建了某个类的对象，然后让浏览器显示这个对象的信息，这时可以将类定义为VO类：因为不希望也不允许浏览器修改对象的属性值。VO类一般作为Java成员内部类或者静态内部类使用。Java的record关键字是定义VO类的有力工具。

FO类是完成一定特定功能的Java类，如完成网络通信的Java类、进行数据库操作的Java类、进行业务处理的Java类等。不同的FO类还被赋予了不同的名称，如专门完成数据库操作的类被赋予了DAO（data access object）类名称等。

5.5.3 任务实施

1. PO类

下面以书籍对象为例介绍PO类的典型模式。对于一本书籍，显然都包括以下属性：书籍名称name、书籍价格price、书籍出版社press、书籍简介memo、书籍作者author等。对于每个属性，都需要定义相应的设置setter( )及读取getter( )方法。为此，在ch05-01工程下新建名为com.ttt.best的Java包，在这个包下新建Book类。IDEA为定义PO类提供了便利的工具：可以为PO类自动生成setter( )、getter( )方法及构造函数等。如图5-5所示，在Book代码编辑区的任一空白区域右击，在弹出的快捷菜单中选择Generate命令，之后在再次弹出的窗口中选择希望自动生成的代码，如图5-6所示。

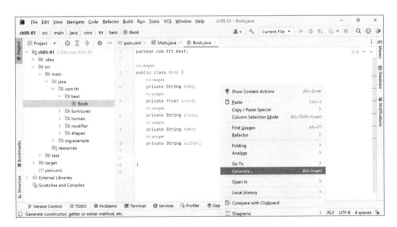

图5-5　IDEA为PO类自动生成setter、getter方法及构造函数

图5-6　选择希望自动生成的代码

IDEA会根据选择自动生成需要的代码。例如，可以为Book类自动生成构造函数、setter( )方法、getter( )方法等。Book类的代码如下（源代码为05-15.java）：

```java
package com.ttt.best;
public class Book {
    private String name;
    private float price;
    private String press;
    private String memo;
    private String author;
    public Book(String name, float price, String press, String memo,
    String author) {
        this.name = name;
        this.price = price;
        this.press = press;
        this.memo = memo;
        this.author = author;
    }
    public String getName() {
        return name;
    }
    public void setName(String name) {
        this.name = name;
    }
    public float getPrice() {
        return price;
    }
    public void setPrice(float price) {
        this.price = price;
    }
    public String getPress() {
        return press;
    }
    public void setPress(String press) {
        this.press = press;
    }
    public String getMemo() {
        return memo;
    }
    public void setMemo(String memo) {
        this.memo = memo;
    }
    public String getAuthor() {
        return author;
    }
    public void setAuthor(String author) {
        this.author = author;
    }
```

```
    @Override
    public String toString() {
        return "Book{" +
                "name='" + name + '\'' +
                ", price=" + price +
                ", press='" + press + '\'' +
                ", memo='" + memo + '\'' +
                ", author='" + author + '\'' +
                '}';
    }
}
```

2. VO 类

下面通过一个例子来说明 VO 类应用。GrandCreator 类根据用户的不同输入信息创建不同的 VO 类对象（源代码为 05-16.java）。

```
package com.ttt.best;
import java.util.Scanner;
public class GrandCreator {
    public void generate() {
        Scanner sc = new Scanner(System.in);
        System.out.print("Which VO to create: ");
        int which = sc.nextInt();
        switch (which) {
            case 1 -> {
                Sky sky = new Sky("Blue", 10000);
                System.out.println(sky);
            }
            case 2 -> {
                Earth earth = new Earth("Home", 100);
                System.out.println(earth);
            }
            case 3 -> {
                Ocean ocean = new Ocean("tuna", 2000);
                System.out.println(ocean);
            }
            default -> System.out.println("Invalid");
        }
    }
    private record Sky(String color, int height){}
    private record Earth(String name, float area){}
    private record Ocean(String fish, float depth){}
}
```

在 GrandCreator 类中定义了三个私有的 VO 类：Sky 类、Earth 类和 Ocean 类。

```
private record Sky(String color, int height){}
private record Earth(String name, float area){}
private record Ocean(String fish, float depth){}
```

在 GrandCreator 的 generate( ) 方法中，根据用户的输入创建不同的 VO 类对象并显示 VO 对象的信息。

现在修改 Main 类，在 main( ) 方法中创建 GrandCreator 类的对象，并调用 generate( ) 方法创建不同的 VO 对象。Main 类的代码如下（源代码为 05-17.java）：

```
package org.example;
import com.ttt.best.GrandCreator;
public class Main {
    public static void main(String[] args) {
        GrandCreator gc = new GrandCreator();
        gc.generate();
    }
}
```

运行这个程序，显示以下信息。

```
Which VO to create: 1
Sky[color=Blue, height=10000]
```

3. FO 类

关于 FO 类，将在后续章节介绍。

5.6 练习：打印自定义图形形状

编写一个 Java 程序，首先定义一个父类 Shape，这个类包含 name 属性和两个抽象方法：一个方法为 draw( )，用于在屏幕上输出图形的外观形状；另一个方法为 info( )，用于显示图形形状的信息。然后定义这个类的子类：正方形，这个子类重写父类的 draw( ) 方法和 info( ) 方法。最后编写一个 Main 类测试代码，用于验证代码的正确性。提示：在屏幕上输出各个图形的形状时，可用一个"*"代表一个点。

第 6 章 接　　口

在编程实践中经常会遇到这样的场景：两种完全不相关的类却具有相同的方法。对于这种情况该如何处理呢？一种可能的解决办法是：为这两种完全不相关的类定义一个抽象的父类，暂且将这个类命名为 A，在父类 A 中定义这两个类中相同的方法并将它们设置为 abstract 抽象的方法。但是，这种做法存在一个潜在问题：由于 Java 不支持多父类继承，因此，如果这两个类本身都需要从某个父类继承，那么，将不能再继承这个仅包含抽象方法的父类 A。为了解决这个问题，Java 提供了一种新的技术手段，也就是接口：通过接口可以定义不相关类的相同行为，也可以指明类必须要实现的方法。在 Java 编程实践中，接口的应用非常丰富。

6.1　接口及其应用

通过一个例子来介绍接口的概念及其应用。

在数学上有一类非常重要的数：复数。从中学数学知识可以知道，任何一个复数都可以表示为 a+b*i 的形式，其中，a、b 为实数，i 为虚数单位且 $i^2=-1$。复数是支持加法、减法运算的，同时，可以将复数增大或者缩小。在数学上还有一类非常重要的数：向量。从中学数学知识可以知道，n 维向量是具有 n 个元素的一列或者一行实数，一般可表示为"$(a_1,a_2,a_3,\cdots,a_n)$"的形式，其中，$a_1, a_2, \cdots, a_n$ 为实数。向量也是支持加法、减法运算的，同时，可以将向量增大或者缩小。

复数和向量这两类数学类的形式和表达的含义各不相同，但是它们都支持相同的运算：加法和减法。如何定义共有的方法呢？可以使用接口来解决这个问题。为了编写和测试本章的例子，在 IDEA 中新建一个名为 ch06-01 的工程，并在该工程中新建名为 com.ttt.math 的包。新建的 ch06-01 工程如图 6-1 所示。

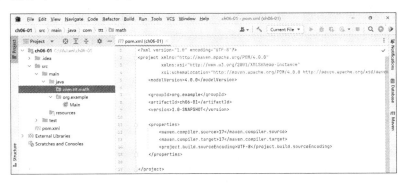

图 6-1　新建的 ch06-01 工程

6.1.1 定义接口

为了给复数类和向量类定义共有的方法，可以定义一个名为 ICompute 的接口。注意，接口名称的命名一般采用"I+动词"的形式。为此，在 ch06-01 的包下新建名为 ICompute 的接口：选择 com.ttt.math → New → Java Class 命令，在弹出的窗口中选择 Interface，并在输入框中输入 ICompute，如图 6-2 所示。

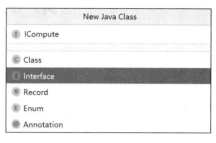

图 6-2　新建 Interface 接口

在 ICompute 接口中定义进行加法运算的 add( ) 方法、进行减法运算的 subtract( ) 方法、进行倍增运算的 enlarge( ) 方法、进行收缩运算的 shrink( ) 方法。ICompute 接口的完整代码如下所示（源代码为 06-01.java）：

```
package com.ttt.math;
public interface ICompute {
    public abstract void add(float... args);
    public abstract void subtract(float... args);
    public abstract void enlarge(int factor);
    public abstract void shrink(int factor);
}
```

由于 Java 的接口只定义方法的原型，不进行方法的具体实现，因此，在 ICompute 接口中只定义了四个方法的名称、参数、返回值等方法的原型，并没有方法的具体实现代码。接口中方法的具体实现代码由实现接口的类完成。

在 ICompue 接口中定义的两个方法 add( ) 和 subtract( ) 用到了可变参数：方法参数的个数在调用时是可变的。

```
void add(float... args);
void subtract(float... args);
```

本质上 Java 的可变参数是基于数组实现的，因此，可以用使用数组的方式使用可变参数，也就是可以使用 args.length 获得参数的个数，可以使用"args[下标]"方式访问每个参数的值。

由于接口中的默认方法都是由 public 和 abstract 修饰的，因此，在接口的定义中可以省略方法前面的 public 和 abstract 关键字，因此，以上 ICompute 接口的代码可以简写为以下代码（源代码为 06-02.java）。

```
package com.ttt.math;
public interface ICompute {
```

```
    void add(float... args);
    void subtract(float... args);
    void enlarge(int factor);
    void shrink(int factor);
}
```

Java 中，定义接口的一般形式如下：

```
public interface 接口名称 {
    // 接口的属性
    // 接口的方法
}
```

在接口中也可以定义属性：Java 规定，接口中的所有属性默认都是由 public、static、final 修饰的，也就是说，接口中定义的所有属性都是公开的、静态的和不可修改的。不可以为接口定义构造函数。

6.1.2 实现接口

复数类和向量类虽然是两个完全不同的类，但是它们具有 ICompue 接口中定义的方法，因此，可以在定义复数类和向量类时，指明这两个类必须实现 ICompute 接口。先定义复数（Complex）类。为此，在 ch06-01 工程的 com.ttt.math 包下新建名为 Complex 的类。Complex 类代码如下（源代码为 06-03.java）：

```
package com.ttt.math;
public class Complex implements ICompute {
    private float a, b;
    public Complex(float a, float b) {
        this.a = a;
        this.b = b;
    }
    @Override
    public String toString() {
        return "Complex{" +
                "a=" + a +
                ", b=" + b +
                '}';
    }
    @Override
    public void add(float... args) {
        if (args.length != 2) return;
        a = a + args[0];
```

```java
        b = b + args[1];
    }
    @Override
    public void subtract(float... args) {
        if (args.length != 2) return;
        a = a - args[0];
        b = b - args[1];
    }
    @Override
    public void enlarge(int factor) {
        a = a*factor;
        b = b*factor;
    }
    @Override
    public void shrink(int factor) {
        a = a/factor;
        b = b/factor;
    }
}
```

使用语句:

```java
public class Complex implements ICompute {
```

规定所定义的 Complex 类必须实现 ICompute 接口。如果指明一个类实现某个接口，那么，这个类要么实现这个接口中的所有方法，要么使用 abstract 关键字修改这个类，以表明这个类是抽象的。

类似地，可以实现向量（Vector）类。为此，在 ch06-01 工程的 com.ttt.math 包下新建名为 Vector 的类。Vector 类代码如下（源代码为 06-04.java）:

```java
package com.ttt.math;
import java.util.Arrays;
public class Vector implements ICompute {
    private float[] element;
    public Vector(float... element) {
        this.element = element;
    }
    @Override
    public String toString() {
        return "Vector{" +
                "element=" + Arrays.toString(element) +
                '}';
    }
```

```java
@Override
public void add(float... args) {
    if (element.length != args.length) return;
    for(int i=0; i<element.length; i++) {
        element[i] += args[i];
    }
}
@Override
public void subtract(float... args) {
    if (element.length != args.length) return;
    for(int i=0; i<element.length; i++) {
        element[i] -= args[i];
    }
}
@Override
public void enlarge(int factor) {
    for(int i=0; i<element.length; i++) {
        element[i] *= factor;
    }
}
@Override
public void shrink(int factor) {
    for(int i=0; i<element.length; i++) {
        element[i] /= factor;
    }
}
}
```

Vector 类的定义与 Complex 类的定义非常类似，不再赘述。只是在 toString( ) 方法中用到了一个在 JDK 中定义的类 Arrays。

```java
"element=" + Arrays.toString(element) +
```

在 Arrays 类中的 toString( ) 静态方法的功能是将数组的元素转换成字符串。

6.1.3 使用接口及 instanceof 关键字在接口中的应用

现在可以测试 ICompute 接口、Complex 类和 Vector 类的正确性。可以直接创建 Complex 类及 Vector 类的对象，进行加法、减法、倍增、收缩操作，然后显示结果值。为此，修改 Main 类，在 main( ) 函数中完成这些操作。修改后的 Main 类代码如下（源代码为 06-05.java）：

```java
package org.example;
import com.ttt.math.Complex;
import com.ttt.math.Vector;
public class Main {
    public static void main(String[] args) {
        Complex c = new Complex(10, 20);
        c.add(100, 200);
        System.out.println(c.toString());
        Vector v = new Vector(10, 20, 30, 40);
        v.shrink(10);
        v.add(100, 200, 300, 400);
        System.out.println(v.toString());
    }
}
```

无须对 Main 类做过多解释。运行这个程序，显示以下结果。

```
Complex{a=110.0, b=220.0}
Vector{element=[101.0, 202.0, 303.0, 404.0]}
```

在 Java 中，不能创建接口对象，但是可以定义接口变量并使接口变量引用接口的对象。为了说明这个现象，修改 Main 类的代码，在 main( ) 函数中定义 ICompute 接口变量 compute，并使接口变量引用 Complex 类和 Vector 类的对象。修改后的 Main 类的代码如下（源代码为 06-06.java）：

```java
package org.example;
import com.ttt.math.Complex;
import com.ttt.math.ICompute;
import com.ttt.math.Vector;
import java.awt.desktop.SystemEventListener;
public class Main {
    public static void main(String[] args) {
        ICompute compute;
        compute = new Complex(10, 20);
        compute.subtract(10, 10);
        System.out.println(compute instanceof Complex);
        System.out.println(compute instanceof Vector);
        compute = new Vector(10, 20, 30, 40, 50);
        compute.enlarge(2);
        System.out.println(compute instanceof Complex);
        System.out.println(compute instanceof Vector);
    }
}
```

该程序首先定义了 ICompute 接口的变量,并分别使用语句:

```
compute = new Complex(10, 20);
```

和

```
compute = new Vector(10, 20, 30, 40, 50);
```

使 compute 变量引用接口的对象,然后使用 instanceof 关键字判断 compute 变量当前所引用对象的类型。

```
System.out.println(compute instanceof Complex);
System.out.println(compute instanceof Vector);
```

运行这个程序,显示以下结果。

```
true
false
false
true
```

6.1.4 接口的继承

接口是可以被继承的:可以基于已经存在的接口作为父接口进一步定义新的子接口。例如,对于 Complex 类和 Vector 类,除了可以进行加法、减法、倍增、收缩运算外,还可以进行比较运算,即比较两个对象是否相等。为此,可以定义一个新的接口 ICompute2,这个接口继承 ICompute 接口。ICompute2 接口的代码如下(源代码为 06-07.java):

```
package com.ttt.math;
public interface ICompute2 extends ICompute {
    public abstract boolean compare(float... args);
}
```

接口的继承使用与类的继承一致的关键字 extends。定义接口继承的一般形式如下:

```
访问限定符 interface 新接口名称 extends 父接口名称 {
    // 接口属性
    // 接口方法
}
```

对于 ICompute2 接口,该接口继承自 ICompute 接口,并在该接口中新增了 compare( )

方法。

```
public abstract boolean compare(float... args);
```

当然，可以修改上面已经定义的 Complex 类和 Vector 类实现这个子接口。此处不再赘述。

在 IDEA 中创建接口及实现接口

6.2 接口的默认方法、静态方法和私有方法

如前所述，在接口中定义的所有属性默认都由 public、static、final 修饰符进行修饰，接口中的所有方法默认都由 public、abstract 修饰符进行修饰。随着 Java 语言的进步，Java 允许在接口中定义默认方法、静态方法和私有方法。

首先介绍接口的默认方法。考虑这样一种应用场景：如果接口中定义了多个方法，其中某个方法对所有实现了这个接口的类的逻辑都是一致的，那么，就没有必要在每个类中都重复实现这个方法，而是可以将这个方法的实现直接放置在接口的定义中。实现了这个接口的任何类将自动继承这个默认方法，同时，实现这个接口的类还可以根据需要重写接口的默认方法。定义接口默认方法的一般形式如下：

```
访问限定符 interface 接口名 {
    // 接口的属性
    // 接口的抽象方法
    public default 返回值 方法名 ( 参数 ) {
        // 默认方法的实现代码
    }
}
```

接口中的方法在默认情况下都是公开的，因此，在定义接口默认方法时，public 关键字是可以省略的。

现在介绍接口的静态方法。接口的静态方法与接口的默认方法类似，只要将定义接口默认方法的 default 关键字换成 static 关键字即可。与类的静态方法类似，接口的静态方法属于接口，实现包含静态接口方法的类不可重写接口的静态方法，并且接口的静态方法也不会被实现接口的类继承。定义接口静态方法的一般形式如下：

```
访问限定符 interface 接口名 {
    // 接口的属性
    // 接口的方法
    public static 返回值 方法名 ( 参数 ) {
        // 静态方法的实现代码
    }
}
```

访问接口的静态方法的一般形式如下：

```
接口的名称 . 接口静态方法 ( 参数 ) ;
```

接口中的方法在默认情况下都是公开的，因此，在定义接口静态方法时，public 关键字是可以省略的。

最后介绍接口的私有方法。接口的私有方法，顾名思义，就是在接口中定义的只能被接口的默认方法使用的且被 private 访问限定符修饰的方法。接口私有方法的一般使用场景是将接口的默认方法中的共性功能抽取出来，并将这些共性功能定义为接口的私有方法，供默认方法调用。定义接口私有方法的一般形式如下：

```
访问限定符 interface 接口名 {
    // 接口的属性
    // 接口的抽象方法
    private 返回值 方法名 ( 参数 ) {
        // 私有方法的实现代码
    }
}
```

下面举例说明接口的默认方法、静态方法和私有方法的定义和使用。定义一个名为 MyInterface 的接口，该接口中包含默认方法、静态方法和私有方法。然后定义一个实现该接口的类 MyExample 类。为了保持 ch06-01 工程的逻辑清晰，在 ch06-01 工程中新建一个名为 com.ttt.notice 的包，将 MyInterface 接口和 MyExample 类放置在该包下。MyInterface 接口的代码如下（源代码为 06-08.java）：

```java
package com.ttt.notice;
public interface IMyInterface {
    public abstract void ordinary();
    public static void display() {
        System.out.println("这是静态方法的输出");
    }
    public default void info() {
        System.out.println("这是默认方法的输出");
        show();
    }
    private void show() {
        System.out.println("这是私有方法的输出");
    }
}
```

在 IMyInterface 接口中，定义了普通方法 ordinary( )、静态方法 display( )、默认方法 info( )、私有方法 show( )。注意接口的私有方法只能在接口的默认方法中调用。现在编写 MyExample 类。MyExample 类的代码如下（源代码为 06-09.java）：

```
package com.ttt.notice;
public class MyExample implements IMyInterface {
    @Override
    public void ordinary() {
        System.out.println(" 这是普通方法的输出 ");
    }
    @Override
    public void info() {
        System.out.println(" 可以重写接口的默认方法 " +
            " 也可以不重写而直接使用接口中的默认方法 ");
        System.out.println(" 这是重写后的默认方法 info() 的输出 ");
    }
}
```

MyExample 类实现了 IMyInterface 接口，因此，需要实现接口的普通方法 ordinary( )。当然，MyExample 类可以根据需要选择是否重写接口的默认方法 info( )，在这里，重写了接口的默认方法。

现在修改 Main 类，在 Main 类的 main( ) 方法中创建 MyExample 类的对象，并调用相应方法显示信息。Main 类的代码如下（源代码为 06-10.java）：

```
package org.example;
import com.ttt.notice.IMyInterface;
import com.ttt.notice.MyExample;
public class Main {
    public static void main(String[] args) {
        MyExample me = new MyExample();
        me.ordinary();
        me.info();
        IMyInterface.display();
    }
}
```

注意其中的语句：

```
IMyInterface.display();
```

直接使用接口名称调用接口的静态方法。运行这个程序，显示如下信息。

```
这是普通方法的输出
可以重写接口的默认方法，也可以不重写而直接使用接口中的默认方法
这是重写后的默认方法 info() 的输出
这是静态方法的输出
```

6.3 函数式接口和 lambda 表达式

在 Java 的编程实践中，函数式接口非常有用也非常常见，它有且只有一个抽象函数。它可以包含其他类型的方法，如默认方法、静态方法、私有方法。这节对函数式接口和 lambda 表达式进行介绍。

6.3.1 函数式接口

为了明确地告知 Java 编译器某个接口是函数式接口，Java 为这类接口提供了一个特定的注解：@FunctionalInterface，这个注解用于接口的定义上。下面通过一个例子说明函数式接口的使用。

在 ch06-01 工程下新建一个名为 com.ttt.function 的程序包，在这个包下新建一个名为 IMyFunction 的函数式接口。新建完成的 IDEA 工程界面如图 6-3 所示。

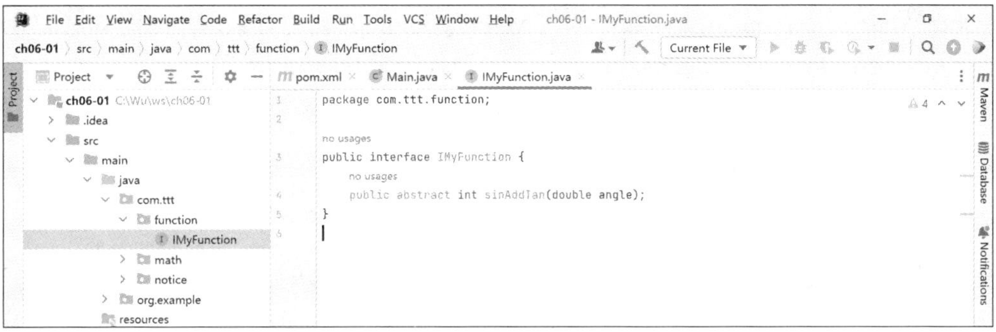

图 6-3 新建完成的 IDEA 工程界面

IMyFunction 函数式接口的代码如下（源代码为 06-11.java）：

```
package com.ttt.function;
@FunctionalInterface
public interface IMyFunction {
    public abstract double sinAddTan(double angle);
}
```

通过 @FunctionalInterface 注解（也称为 Java 标注），Java 编译器在编译时会对接口进行检查，确保该接口满足函数式接口的要求。在 Java 中定义函数式接口的一般形式如下：

```
@FunctionalInterface
访问限定符 interface 接口名称 {
    // 接口的属性
    // 接口的静态方法、默认方法、私有方法定义
```

```
    // 接口的抽象方法
}
```

一旦定义了函数式接口,可以定义 Java 类像实现其他普通接口一样对函数式接口进行实现。例如,下面的 MyFunctionImpl 类实现类 IMyFunction 接口(源代码为 06-12.java)。

```
package com.ttt.function;
public class MyFunctionImpl implements IMyFunction {
    @Override
    public double sinAddTan(double angle) {
        double x = Math.sin(angle) + Math.tan(angle);
        System.out.println("这个幅度的 sin 值+tan 值是: " + x);
        return x;
    }
}
```

修改 Main 类,在 main() 函数中创建 MyFunctionImpl 类的对象,并调用 sinAddTan() 方法显示一个幅度的值。Main 类的代码如下(源代码为 06-13.java):

```
package org.example;
import com.ttt.function.MyFunctionImpl;
public class Main {
    public static void main(String[] args) {
        MyFunctionImpl mfi = new MyFunctionImpl();
        double t = mfi.sinAddTan(20);
        System.out.println(t);
    }
}
```

运行这个程序,显示以下信息。

```
这个幅度的 sin 值+tan 值是: 3.15010619495237
3.15010619495237
```

6.3.2 使用匿名内部类实现接口

还可以使用匿名内部类实现接口。

匿名内部类是 Java 内部类的一种。匿名内部类适用于仅使用一个局部类一次的场景。例如,如果第 6.3.1 小节定义的 Java 类 MyFunctionImpl 仅会在 Main 类中使用一次,那么,就没必要定义这个专门的 Java 类,而是在需要的地方直接使用匿名内部类即可完成同样的功能。为此,修改 Main 类为以下代码,可以实现同样的功能(源代码为 06-14.java)。

```java
package org.example;
import com.ttt.function.IMyFunction;
public class Main {
    public static void main(String[] args) {
        // 下面的代码定义了实现 IMyFunction 接口的匿名内部类并创建了这个类的对象
        IMyFunction im = new IMyFunction() {
            @Override
            public double sinAddTan(double angle) {
                double x = Math.sin(angle) + Math.tan(angle);
                System.out.println("这个幅度的 sin 值 +tan 值是: " + x);
                return x;
            }
        };
        double t = im.sinAddTan(20);
        System.out.println(t);
    }
}
```

使用:

```java
IMyFunction im = new IMyFunction() {
    @Override
    public double sinAddTan(double angle) {
        double x = Math.sin(angle) + Math.tan(angle);
        System.out.println("这个幅度的 sin 值 +tan 值是: " + x);
        return x;
    }
};
```

这段代码直接创建了实现 IMyFunction 接口的匿名内部类，创建了这个匿名内部类的对象，并使用变量 im 引用这个对象。之后，使用以下代码：

```java
double t = im.sinAddTan(20);
System.out.println(t);
```

调用这个对象的方法并显示信息。运行这个程序，显示与第 6.3.1 小节例子同样的结果。注意，匿名内部类可以实现任何接口，而不只限于实现函数式接口。

Java 还提供了实现函数式接口的更为优雅也更为专业的方式：lambda 表达式。

6.3.3 lambda 入门：使用 lambda 表达式实现函数式接口

现在，使用 lambda 表达式修改第 6.3.2 小节的 Main 类。修改后的 Main 类代码如下（源代码为 06-15.java）：

```
package org.example;
import com.ttt.function.IMyFunction;
public class Main {
    public static void main(String[] args) {
        IMyFunction im = (angle)-> {
            double x = Math.sin(angle) + Math.tan(angle);
            System.out.println("这个幅度的 sin 值+tan 值是: " + x);
            return x;
        };
        double t = im.sinAddTan(20);
        System.out.println(t);
    }
}
```

这段代码，首先定义了一个 IMyFunction 函数式接口的变量 im，然后，通过 lambda 表达式的特有语法创建了一个 lambda 表达式。

本质上，lambda 表达式就是对函数式接口中抽象方法的隐式实现。观察一下 IMyFunction 函数式接口可知，在 IMyFunction 函数式接口中只有一个抽象方法。

```
public abstract double sinAddTan(double angle);
```

因此，这个 lambda 表达式就是实现函数式接口的 sinAddTan(double angle) 方法：lambda 表达式括号及其中的参数就对应 sinAddTan(double angle) 抽象函数中的参数，lambda 表达式中用大括号括起来的代码就是对抽象函数的实现代码。

6.3.4 lambda 表达式基本语法

从第 6.3.3 小节对 lambda 表达式的入门介绍可以知道，lambda 表达式的一般形式如下：

```
(参数列表) -> {
    代码块;
}
```

参数说明如下。

(1) 参数列表：参数列表是函数式接口中对应接口抽象方法的参数，当只有一个参数时可以省略圆括号。参数类型可以明确声明也可不声明，而且参数类型可以省略：如果需要省略，由 Java 根据接口中的方法定义自动推断，每个参数的类型都要省略。

(2) ->：lambda 表达式的符号，可以理解为"函数式接口中抽象方法的具体实现代码由如下代码块定义"。

(3) 代码块：由一系列程序语句组成，既可以是表达式也可以代码块，是函数式接口中抽象方法的实现代码，等同于方法的方法体。如果代码块中只有一条语句，则大括号"{ }"可以被省略；如果是一条表达式语句，在省略大括号的同时，还可以将表达式语句写在"–>"的后面，不需要使用 return 语句。

6.3.5 接口方法引用

使用接口方法引用可以使函数式接口变量直接访问某个类或者实例已经存在的方法。接口方法的引用主要应用在以下场景：如果某个类已经定义了满足函数式接口的方法，那么，可以使函数式接口变量直接指向这个类的实现方法，而不需要自己重新编写函数式接口方法的代码。这段文字比较抽象，理解起来比较困难，通过一个具体例子来说明接口方法引用的应用。

接口方法引用举例

6.4 接口、匿名内部类和 lambda 表达式应用实践

引入接口技术机制的初衷是规范类的行为：类所封装的对象可以是完全不同的，但是，它们可以具有相似的行为。接口就是用来规范类的行为的：可以让不同的类实现相同的接口，从而使类具备相似行为。

接口是 Java 实现模块化程序设计的重要支撑。具体来说就是，在进行系统设计时，首先设计好系统应该具备的行为，也就是定义好一组接口，然后编码人员通过类实现这些接口，进而完成系统的编码。在极端情况下，假设某个编码人员的代码出现问题，并且由于该编码人员的编程逻辑混乱，难以对其代码进行修正，这时完全可以安排另一个编码人员根据所定义的接口重新编写这部分代码，并替换原来的错误代码。这种替换是不会影响系统其他部分代码功能的，接口的重要意义正在于此。这种思想被大量应用到 Java 的框架设计中。例如，著名的 Spring 框架就应用了这种思想。

一旦定义了接口，就可以采用多种方法对接口进行实现：可以采用独立的 Java 类实现接口；可以采用匿名内部类实现接口；对于函数式接口，可以使用 lambda 表达式实现接口等。至于如何选择实现机制，基本原则是：如果接口比较简单，特别是函数式接口，则建议使用匿名内部类或者 lambda 表达式实现接口；如果接口比较复杂，还是建议使用独立类的方式实现接口。

6.5 案例：按价格排序不同产品

通过一个综合案例结束本章内容的介绍，该案例对不同类型的食品按照其价格进行排序并显示。这个案例涉及类、类的继承、接口、将接口作为方法参数，并且将 lambda 表达式作为实参进行传递。

6.5.1 案例任务

编写一个 Java 代码，可以比较任何食品的价格高低并进行排序。程序要求首先定义一个函数式接口 IComparable，该接口包含一个抽象方法 compare( )，可对食品的价格高低进行比较，并根据两个食品的价格高低返回 -1、0 和 1。然后定义多个食品类，包括父类 Eating 类及其子类 Cookie、Bread 和 Cake，并创建包含这些子类对象的一个数组，再定义一个可以对数组中的食品对象按价格进行排序的方法。最后，编写一个测试类，用于测试代码的正确性。

6.5.2 任务分析

分析案例任务要求，从中提取出要定义的接口、类及其相应功能要求。基于分析结果可知，需要定义以下接口和类。
（1）IComparable 函数式接口。
（2）食品父类 Eating 及其子类 Cookie、Bread 和 Cake。
（3）Using 类，创建食品数组，定义排序方法。
（4）Main 类，测试程序的正确性。

6.5.3 任务实施

为此，在 ch06-01 的工程下新建 com.ttt.food 包，在包下新建 IComparator 接口和 Bread 类、Cake 类、Cookie 类、Eating 类、Using 类，如图 6-4 所示。

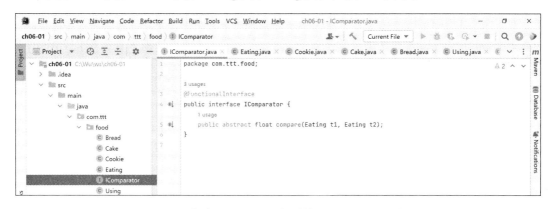

图 6-4 新建 com.ttt.food 包后的 ch06-01 工程结构

IComparator 接口的代码如下，显然，IComparator 接口是一个函数式接口（源代码为 06-16.java）。

```
package com.ttt.food;
```

```java
@FunctionalInterface
public interface IComparator {
    public abstract float compare(Eating t1, Eating t2);
}
```

Eating 类的代码如下（源代码为 06-17.java）：

```java
package com.ttt.food;
public class Eating {
    private String name;
    private float price;
    private String producer;
    public Eating(String name, float price, String producer) {
        this.name = name;
        this.price = price;
        this.producer = producer;
    }
    public String getName() {
        return name;
    }
    public float getPrice() {
        return price;
    }
    public String getProducer() {
        return producer;
    }
}
```

Eating 类包含三个属性、一个构造函数和三个 getter( ) 方法，这个类非常简单，无须赘述。下面定义 Eating 类的三个子类：Cake 类、Bread 类和 Cookie 类。这三个子类都非常简单，无须赘述。Cake 类的代码如下（源代码为 06-18.java）：

```java
package com.ttt.food;
public class Cake extends Eating {
    private String ingredient;   // 蛋糕的主要成分
    private int calorie;         // 能量大小
    public Cake(String name, float price, String producer, String ingredient, int calorie) {
        super(name, price, producer);
        this.ingredient = ingredient;
        this.calorie = calorie;
    }
    public String getIngredient() {
        return ingredient;
```

```java
        }
        public int getCalorie() {
            return calorie;
        }
        @Override
        public String toString() {
            return "Cake{" +
                    "name=" + super.getName() +
                    ", price=" + super.getPrice() +
                    ", producer=" + super.getProducer() +
                    ", ingredient='" + ingredient + '\'' +
                    ", calorie=" + calorie +
                    '}';
        }
}
```

Bread 类的代码如下（源代码为 06-19.java）：

```java
package com.ttt.food;
public class Bread extends Eating {
    private String date;     // 面包的生产日期
    public Bread(String name, float price, String producer, String date) {
        super(name, price, producer);
        this.date = date;
    }
    public String getDate() {
        return date;
    }
    @Override
    public String toString() {
        return "Bread{" +
                "name=" + super.getName() +
                ", price=" + super.getPrice() +
                ", producer=" + super.getProducer() +
                ", date='" + date + '\'' +
                '}';
    }
}
```

Cookie 类的代码如下（源代码为 06-20.java）：

```java
package com.ttt.food;
public class Cookie extends Eating {
    private String shape;     // 饼干的外形
```

```
    public Cookie(String name, float price, String producer, String shape) {
        super(name, price, producer);
        this.shape = shape;
    }
    public String getShape() {
        return shape;
    }
    @Override
    public String toString() {
        return "Cookie{" +
                "name=" + super.getName() +
                ", price=" + super.getPrice() +
                ", producer=" + super.getProducer() +
                ", shape=" + shape +
                '}';
    }
}
```

现在编写 Using 类。这个类的主要作用是创建食品类数组，并定义一个方法，根据所给定的 IComparator 接口对食品数组中的食品进行排序。Using 类的代码如下（源代码为 06-21.java）：

```
package com.ttt.food;
import java.util.Random;
public class Using {
    private Eating[] foods;
    public Using(int n) {
        Random random = new Random();
        foods = new Eating[n];
        for(int i=0; i<n; i++) {
            foods[i] = switch (random.nextInt(3)) {
                case 0 -> new Cake("Cake"+i, random.nextFloat(50),
                        "P"+i, "Rice", random.nextInt(100));
                case 1 -> new Bread( "Bread" +i, random.nextFloat(20),
                        "P"+i, "2023年09月20日");
                case 2 -> new Cookie("Cookie"+i, random.nextFloat(30),
                        "P"+i, "圆形");
                default -> null;
            };
        }
    }
    public void sort(IComparator comparator) {
        for(int i=0; i<foods.length-1; i++) {
            for(int j=i+1; j<foods.length; j++) {
```

```
                float r = comparator.compare(foods[i], foods[j]);
                if (r > 0) {
                    Eating e = foods[i];
                    foods[i] = foods[j];
                    foods[j] = e;
                }
            }
        }
    }
    public void show() {
        for(Eating e : foods)
            System.out.println(e.toString());
    }
}
```

注意其中的 sort( ) 方法。

```
public void sort(IComparator comparator) {
    for(int i=0; i<foods.length-1; i++) {
        for(int j=i+1; j<foods.length; j++) {
            float r = comparator.compare(foods[i], foods[j]);
            if (r > 0) {
                Eating e = foods[i];
                foods[i] = foods[j];
                foods[j] = e;
            }
        }
    }
}
```

这个方法有一个 IComparator 类型的参数。将某个接口作为方法的参数是 Java 编程中非常重要的技巧。在该方法中，使用语句：

```
float r = comparator.compare(foods[i], foods[j]);
```

对食品进行比较，并根据比较结果对食品对象进行排序。这里使用了最简单的冒泡排序对食品进行升序排序：

```
if (r > 0) {
    Eating e = foods[i];
    foods[i] = foods[j];
    foods[j] = e;
}
```

最后，修改 Main 类，创建 Using 类的对象，并调用其中的 sort( ) 方法对食品进行排序。Main 类的代码如下（源代码为 06-22.java）：

```java
package org.example;
import com.ttt.food.IComparator;
import com.ttt.food.Using;
public class Main {
    public static void main(String[] args) {
        Using u = new Using(5);
        IComparator lambda = (e1, e2) -> (e1.getPrice() - e2.getPrice());
        u.sort(lambda);
        u.show();
    }
}
```

注意其中的语句：

```
IComparator lambda = (e1, e2) -> (e1.getPrice() - e2.getPrice());
u.sort(lambda);
```

这里使用 lambda 表达式实现类 IComparator 函数式接口，并将它作为参数传递给 sort( ) 方法。运行这个程序，显示以下结果。

```
Bread{name=Bread0, price=2.331642, producer=P0, date='2023年09月20日'}
Cookie{name=Cookie2, price=5.1906557, producer=P2, shape=圆形}
Cake{name=Cake1, price=10.386335, producer=P1, ingredient='Rice', calorie=54}
Bread{name=Bread4, price=15.738033, producer=P4, date='2023年09月20日'}
Cake{name=Cake3, price=41.081627, producer=P3, ingredient='Rice', calorie=22}
```

6.6　练习：计算空间中两点的距离

编写一个 Java 程序，计算空间中任意两点之间的距离。程序要求首先定义一个接口 IDistance，该接口定义了计算空间中任意两点之间距离的方法和一个显示空间任意点对象信息的默认方法，然后定义两种类型的空间点类：二维空间点、三维空间点，这些类都要实现 IDistance 接口，最后编写一个测试类，用于测试代码的正确性。

第 7 章 枚 举 类 型

考虑这样一种应用：定义某种类型的变量，并且这个变量只能被赋予某些指定的值，如果所赋予的值不是事先规定的特定值，那么 Java 编译器就能直接指出这个错误。Java 提供了这种机制，这种机制就是枚举类型。采用枚举类型能克服程序员在编程时的粗心所导致的低级错误，当然，枚举类型能解决的问题远不止于此，如可以使用枚举类型表示程序的状态码或者错误码等。

7.1 枚举类型入门：一个表示四季的枚举类型

一年有四个季节：春、夏、秋、冬。能不能定义一个类表示"春、夏、秋、冬"四季，并且，之后所定义的这个类的变量只能被赋予规定的"春、夏、秋、冬"这四个值呢？答案是可以的。Java 提供的枚举类型能完美地满足这种要求。例如，可以定义一个包含"春、夏、秋、冬"四个取值的枚举类 Season。为此，新建一个名为 ch07-01 的 Java 工程，并在其下新建名为 com.ttt.enums 的程序包，然后新建名为 Season 的类。新建的 ch07-01 工程如图 7-1 所示。

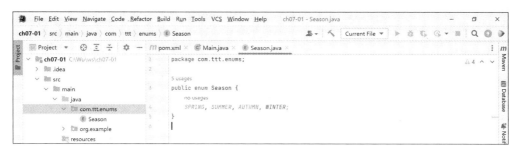

图 7-1 新建的 **ch07-01** 工程

Season 枚举类型的代码如下（源代码为 07-01.java）：

```
package com.ttt.enums;
public enum Season {
    SPRING, SUMMER, AUTUMN, WINTER;
}
```

这里定义了一个简单的枚举类型 Season，对于 Season 类型的变量，可以赋予 SPRING、SUMMER、AUTUMN、WINTER 四个值之一。定义简单枚举类型的一般形式如下：

```
访问限定符 enum 枚举类型名称 {
    // 可用的值，之间用逗号","分隔，最后一个值的后面加上分号";"
}
```

现在修改 Main 类，定义 Season 枚举类型的变量，并显示相关信息。修改后的 Main 类代码如下（源代码为 07-02.java）：

```java
package org.example;
import com.ttt.enums.Season;
public class Main {
    public static void main(String[] args) {
        // 只能为枚举变量赋予预先设定的枚举值，否则编译报错
        Season s1 = Season.WINTER;
        System.out.println(s1.toString());
        System.out.println(s1.ordinal());
        System.out.println(s1.name());
        System.out.println("------------------");
        // 其中的参数只能是枚举值对应的字符串常量或变量，否则运行时错误
        Season s2 = Season.valueOf("SPRING");
        System.out.println(s2.toString());
        System.out.println(s2.ordinal());
        System.out.println(s2.name());
        System.out.println("------------------");
        for(Season s : Season.values())
            System.out.println(s.toString());
    }
}
```

在 main( ) 方法中，使用语句：

```
Season s1 = Season.WINTER;
```

定义了 Season 类型的变量 s1，并使之引用 Season.WINTER 枚举量。然后使用语句：

```
System.out.println(s1);
System.out.println(s1.ordinal());
System.out.println(s1.name());
```

显示变量 s1 的相关信息。这里可能会有疑问：在 Season 枚举类型中并没有定义 toString( ) 方法、ordinal( ) 方法和 name( ) 方法，为什么可以调用这些方法呢？这些方法是从哪里来的呢？原因是：Java 的枚举类型都自动继承自 java.lang.Enum 类。所有枚举实例都可以调用 Enum 类的方法。Enum 的常用方法如表 7-1 所示。

表 7-1 Enum 的常用方法

方　　法	功　　能
String name( )	返回枚举值名称
int ordinal( )	获取枚举成员的索引位置，每个枚举值在枚举类型中都有一个唯一的序号，从 0 开始
Enum valueOf(String name)	将普通字符串转换为枚举实例
int compareTo(Enum another)	比较两个枚举成员在定义时的顺序，也就是比较两个枚举值的序号大小，返回两个枚举值的序号差
String[] values( )	以数组形式返回枚举类型的所有成员

注意语句：

```
Season s2 = Season.valueOf("SPRING");
```

这条语句通过字符串 "SPRING" 转换为枚举类型中与字符串值相等的枚举值。运行这个程序，显示以下信息。

```
WINTER
3
WINTER
------------------
SPRING
0
SPRING
------------------
SPRING
SUMMER
AUTUMN
WINTER
```

7.2 枚举类型进阶

在定义枚举类型的枚举值时，可以为每个枚举值设定需要的属性，也就是说，可以通过为枚举值设定属性进一步地对枚举值进行附加说明。不仅如此，还可以为枚举类型定义方法。下面通过一个例子说明如何为枚举类型的枚举值设定属性和为枚举类型定义方法。

为此，在 com.ttt.enums 包下新建一个名为 WeekDay 的枚举类型，用以表示星期的名称。在定义该枚举类型时，为每个枚举值设定两个属性：code 属性，为每个枚举值设定一个编码；cn 属性，为每个枚举值设定中文名称。WeekDay 枚举类型的源代码如下（源代码为 07-03.java）：

```
package com.ttt.enums;
```

```java
public enum WeekDay {
    Monday(1, "星期一"), Tuesday(2, "星期二"), Wednesday(3, "星期三"),
    Thursday(4, "星期四"), Friday(5, "星期五"), Saturday(6, "星期六"),
    Sunday(7, "星期日");
    private int code;
    private String cn;
    private WeekDay(int code, String cn) {
        this.code = code;
        this.cn = cn;
    }
    public int getCode() {
        return code;
    }
    public String getCn() {
        return cn;
    }
    @Override
    public String toString() {
        return "WeekDay{" +
                "code=" + code +
                ", cn='" + cn + '\'' +
                '}';
    }
}
```

这段代码中，对 WeekDay 枚举类型的每个枚举值都赋予了一个 code 属性和 cn 属性。例如，对于枚举 Monday，通过以下语句为之赋予 code 属性和 cn 属性。

```
Monday(1, "星期一")
```

再者，为了能够给每个枚举值赋予相应属性，Java 规定，必须定义访问控制符为 private 的枚举类型构造函数，为此，为 WeekDay 定义了以下构造函数。

```java
private WeekDay(int code, String cn) {
    this.code = code;
    this.cn = cn;
}
```

为枚举值的对应属性赋值。当然，可以根据需要，定义其他的枚举类型方法。例如，在这段代码中，定义了 getter( ) 方法和 toString( ) 方法。

现在，修改 Main 类，在 main( ) 函数中定义 WeekDay 变量并引用 WeekDay 的枚举值，然后显示相应信息。修改后的 Main 类代码如下（源代码为 07-04.java）：

```java
package org.example;
import com.ttt.enums.WeekDay;
```

```java
public class Main {
    public static void main(String[] args) {
        WeekDay wd1 = WeekDay.Saturday;
        System.out.print(wd1.name() + ", ");
        System.out.print(wd1.ordinal() + ", ");
        System.out.println(wd1.getCn());
        System.out.println("--------------------");
        WeekDay wd2 = WeekDay.valueOf("Thursday");
        System.out.print(wd2.name() + ", ");
        System.out.print(wd2.ordinal() + ", ");
        System.out.println(wd2.getCn());
        System.out.println("--------------------");
        WeekDay wd3 = WeekDay.Saturday;
        System.out.println(wd1.equals(wd2));
        System.out.println(wd1.equals(wd3));
        System.out.println("--------------------");
        for(WeekDay w : WeekDay.values()) {
            System.out.print(w.getCn() + " ");
        }
    }
}
```

运行这个程序，显示以下信息。

```
Saturday, 5, 星期六
--------------------
Thursday, 3, 星期四
--------------------
false
true
--------------------
星期一 星期二 星期三 星期四 星期五 星期六 星期日
```

7.3 枚举类型应用实践

使用枚举类型可以改善程序的可读性，这是枚举类型最常见的应用场景。例如，程序中有一个表示星期的变量，使用枚举类型既改善了程序的可读性，又减少了代码的出错概率；使用枚举类型表示人的性别具有相似的意义。

除此之外，在软件工程化实践中，经常会遇到这样的场景：一个方法需要根据执行结果返回特定的某些值。举个例子，程序中定义了一个电梯类，该类有一个控制电梯开门的方法openDoor( )。这个方法的返回值需要明确告知调用者方法的执行结果。该方法可能的执行结果包括：门被正常打开、门未打开—电机故障、门未打开—左门被异物卡住、门未打开—右门被异物卡住、门未打开—控制皮带未卡紧、门未打开—电梯箱未到位、门未打开—其他未知错误。

处理这个问题的方法之一就是定义 openDoor( ) 方法的返回值为 int 类型，其中，0 表示门被正常打开，1 表示门未打开—电机故障，2 表示门未打开—左门被异物卡住，3 表示门未打开—右门被异物卡住，4 表示门未打开—控制皮带未卡紧，5 表示门未打开—电梯箱未到位，6 表示门未打开—其他未知错误。当然，使用整数值来表示 openDoor( ) 方法的执行结果，需要编程人员记住每个整数值所代表的含义。这增加了编码人员的困难。这种方式是可行的，但不是最专业的。

处理这个问题的更好方式是使用枚举类型：定义一个枚举类型，表示 penDoor( ) 方法的返回值。例如，可以定义名为 ErrorCode 的枚举类型，代码如下（源代码为 07-05.java）：

```java
package com.ttt.enums;
public enum ErrorCode {
    Success(0, "门被正常打开"),
    Failure1(1, "电机故障"), Failure2(2, "左门被异物卡住"),
    Failure3(3, "右门被异物卡住"), Failure4(4, "控制皮带未卡紧"),
    Failure5(5, "电梯箱未到位"), Failure6(6, "其他未知错误");
    private final int code;
    private final String description;
    private ErrorCode(int code, String description) {
        this.code = code;
        this.description = description;
    }
    public int getCode() {
        return code;
    }
    public String getDescription() {
        return description;
    }
    @Override
    public String toString() {
        return "ErrorCode{" +
                "code=" + code +
                ", description='" + description + '\'' +
                '}';
    }
}
```

由于 openDoor( ) 方法的返回值是一个 ErrorCode 的枚举类型值，因此，方法的调用者从返回值中就可以获得方法执行结果的所有信息，而不需要编码者记住每个值的含义。

7.4　案例：员工 Staff 类

本节结合类的定义和枚举类型，介绍如何在类的属性中使用枚举类型限定属性的取值。

7.4.1 案例任务

每个员工都是有性别的，每个员工也都是有工作时间的。定义一个 Staff 类，在表示员工基本信息的同时，要求定义使用枚举类型表示员工的性别和工作时间，如一个星期的哪几天是需要工作的。

7.4.2 任务分析

这个案例要求定义一个 Staff 类表示员工信息，同时，要求使用枚举类型表示员工的性别和工作时间。因此，在定义 Staff 类的同时，需要定义表示性别的枚举类型和表示工作时间的枚举类型。总结起来，该案例任务需要定义以下的枚举类型和类。

（1）定义表示性别的名为 Gender 枚举类型。
（2）定义表示工作时间的名为 WeekDay 枚举类型。
（3）定义表示员工的 Staff 类。

7.4.3 任务实施

在 ch07-01 工程下新建一个名为 com.ttt.staff 的包。在该包下新建 Gender 枚举类型和 Staff 类。由于在第 7.2 节已经定义了 WeekDay 类，这里可以直接沿用这个枚举类型。Gender 枚举类型的代码如下（源代码为 07-06.java）：

```
package com.ttt.staff;
public enum Gender {
    Male, Female;
}
```

Staff 类的代码如下（源代码为 07-07.java）：

```
package com.ttt.staff;
import com.ttt.enums.WeekDay;
public class Staff {
    private String name;
    private Gender gender;
    private float salary;
    private WeekDay[] workdays;
    public Staff(String name, Gender gender, float salary, WeekDay...
workdays) {
        this.name = name;
        this.gender = gender;
        this.salary = salary;
        this.workdays = workdays;
    }
    public String getName() {
```

```
        return name;
    }
    public Gender getGender() {
        return gender;
    }
    public float getSalary() {
        return salary;
    }
    public WeekDay[] getWorkdays() {
        return workdays;
    }
}
```

现在修改 Main 类，修改后的 Main 类的代码如下（源代码为 07-08.java）：

```
package org.example;
import com.ttt.enums.WeekDay;
import com.ttt.staff.Gender;
import com.ttt.staff.Staff;
public class Main {
    public static void main(String[] args) {
        Staff staff1 = new Staff("张三", Gender.Male, 5050.5f,
                WeekDay.Tuesday, WeekDay.Friday, WeekDay.Saturday);
        System.out.print(staff1.getName() + "性别是: ");
        System.out.println(staff1.getGender().toString());
        System.out.println(staff1.getName() + " 工作时间是:");
        for(WeekDay wd : staff1.getWorkdays()) {
            System.out.println(wd);
        }
    }
}
```

运行这个程序，显示以下结果。

```
张三性别是: Male
张三工作时间是:
WeekDay{code=2, cn='星期二'}
WeekDay{code=5, cn='星期五'}
WeekDay{code=6, cn='星期六'}
```

7.5 练习：水果的成熟季节

大家知道，水果都有名称、形状和成熟季节。编写一个 Java 程序，首先设计一个表示水果的类，其中，水果的成熟季节用枚举类型表示。然后编写一个测试类，用于测试程序的正确性。

第 8 章　Java 基础类的使用

使用 Java 语言开发应用程序都需要使用 JDK，这就是为什么在第 1 章的开始部分需要安装 JDK。JDK 中除了包含 Java 编译器、Java 文档工具、Java 监视工具外，还包括非常重要的内容，那就是 Java 定义的程序类，也就是 Java 根据应用的需要已经事先定义了的一组 Java 程序类。据不完全统计，JDK 中定义了 8000 多个程序类。在这一章对 Java 的基础程序类进行介绍。在后续章节中，还会介绍 Java 其他常用类。当然，熟悉 JDK 中程序类非常有效的方法是看 JDK 文档。

8.1　Java 基本类

java.lang 包是 JDK 的一个非常重要的包，这个包下还包括很多子包。在这个包中定义很多 Java 基本类，包括之前用到的 String 类、Math 类等。本节对 Java 的典型基本类的使用进行介绍。

JDK 中的 Java 类及其帮助文档

8.1.1　Object 类

Object 类是 Java 的祖先类，Java 的任何类都是直接或间接地从这个类派生的，也就是说，Java 的任何类要么是 Object 类的子类，要么是 Object 的子类的子类（孙类）。在定义一个类时，如果没有指明父类，那么默认情况下，Object 就是新定义类的父类。Object 类的常用方法如表 8-1 所示。

表 8-1　Object 类的常用方法

方　　法	功　　能
boolean equals( )	判断两个对象是否相等
String toString( )	将对象转换为可显示的字符串。默认情况下，转换的结果是对象所述类的名称和对象在内存中的地址
int hashCode( )	生成对象的 hash 码。hash 码是可以用来代表这个对象的一个 int 类型的整数

下面举一个简单例子说明 Object 类的使用。特别声明：Object 类一般情况下总是作为 Java 的祖先类被使用，但是，也可以直接创建 Object 类的对象。修改 Main 类的代码，在 main( ) 方法中创建 Object 类的对象，并显示相关信息。Main 类的代码如下（源代码为 08-01.java）：

```java
package org.example;
public class Main {
    public static void main(String[] args) {
        Object obj1 = new Object();
        System.out.println(obj1.hashCode());
        System.out.println(obj1.toString());
        Object obj2 = new Object();
        System.out.println(obj2.hashCode());
        System.out.println(obj1.equals(obj2));
    }
}
```

运行这个程序，显示以下结果。

```
1324119927
java.lang.Object@4eec7777
990368553
false
```

8.1.2 基本数据类型的包装类

前面介绍的 Java 基本数据类型，像 int、long、double、float、char、boolean 等，JDK 都为它们定义了对应的 Java 类，称为基本数据类型的包装类。它们对应的包装类分别为：Integer 类、Long 类、Double 类、Float 类、Char 类和 Boolean 类。通过包装类，可以实现对数据的基本操作，如在 Integer 类中就定义了将字符串转换为整数的方法。下面以 Integer 类的使用为例，对 Java 包装类的使用进行介绍。

修改 Main 类的代码，在 main( ) 方法中使用 Integer 类提供的一系列方法。修改后的 Main 类的代码如下（源代码为 08-02.java）：

```java
package org.example;
public class Main {
    public static void main(String[] args) {
        Integer i1 = Integer.valueOf("12345");
        Integer i2 = 100;    //Java 会自动将 int 类型的值包装成 Integer 对象
        System.out.println("i1=" + i1 + ", i2=" + i2);
        // 返回两个整数的较大者
        System.out.println(Integer.max(123, 234));
        // 返回最大的整数值
        System.out.println(Integer.MAX_VALUE);
        // 返回最小的整数值
        System.out.println(Integer.MIN_VALUE);
        // 返回 int 类型值占用的内存字节数
```

```
            System.out.println(Integer.BYTES);
    }
}
```

运行这个程序，显示以下结果。

```
i1=12345, i2=100
234
2147483647
-2147483648
4
```

8.1.3 大数据类

由于 int 类型、long 类型、double 类型所用的内存字节限制，它们所能表示的数据大小范围是有限的。为了能够表示任一大小的数据，Java 提供了 BigInteger 类表示无大小限制的整数，提供了 BigDecimal 类表示无大小精度限制的实数类。下面以 BigDecimal 类的使用为例，介绍大数据类的使用。

修改 Main 类的代码，在 main( ) 方法中创建 BigDecimal 类的对象，然后进行数据的加、减、乘、除运算并显示结果。修改后的 Main 类代码如下（源代码为 08-03.java）：

```
package org.example;
import java.math.BigDecimal;
import java.math.RoundingMode;
public class Main {
    public static void main(String[] args) {
        BigDecimal bd1 = new BigDecimal("23456313456745456456.97346483
            48349539");
        BigDecimal bd2 = new BigDecimal("5674545656456.12338348349876890
            32539");
        BigDecimal bd3 = bd1.add(bd2);
        BigDecimal bd4 = bd1.divide(bd2, 10, RoundingMode.CEILING);
        System.out.println(bd1);
        System.out.println(bd2);
        System.out.println(bd3);
        System.out.println(bd4);
    }
}
```

注意其中的语句：

```
BigDecimal bd4 = bd1.divide(bd2, 10, RoundingMode.CEILING);
```

第二个参数表示当无法除尽时保留小数点后的位数，第三个参数表示舍入方式。运行这个程序，显示以下结果。

```
234563134567454565456.9734648348349539
5674545656456.12338348349876890322539
234563135134909131291 3.0968483183337228032539
413360203.2095452758
```

8.1.4 System 类

System 类封装了计算机系统的相关设备和部件。例如，之前一直使用 System.out.println( ) 方法显示相关数据，其中的 out 封装了计算机系统的显示设备，用于将信息显示出来。通过 System 类，还可以取得系统的日期时间。下面举一个例子介绍 System 类的主要方法和对象的使用。

修改 Main 类的代码，在 main( ) 方法中创建 BigDecimal 类的对象，然后进行数据的加、减、乘、除运算并显示结果。修改后的 Main 类的代码如下（源代码为 08-04.java）:

```java
package org.example;
import java.util.Scanner;
public class Main {
    public static void main(String[] args) {
        System.out.println("Hello World!");
        Scanner sc = new Scanner(System.in);
        System.out.print("输入一个名字: ");
        String name = sc.nextLine();
        System.out.print("输入一个整数: ");
        int number = sc.nextInt();
        System.out.println(number);
        System.out.println(name);
        // 返回自 1970 年 1 月 1 日 0 点到当前时刻的毫秒数
        System.out.println(System.currentTimeMillis());
        // 直接退出程序运行，并返回整数 100 给操作系统
        System.exit(100);
    }
}
```

运行这个程序，显示以下结果。

```
Hello World!
输入一个名字: Bill
输入一个整数: 100
100
Bill
```

```
1673423794147
Process finished with exit code 100
```

注意运行结果中 100 这个数字，这是语句：

System.exit(100);

返回给操作系统的数据。

8.1.5　Math 类

JDK 对常用的数学运算函数进行了代码实现，如数学中的 sin( ) 函数、cos( ) 函数、tan( ) 函数、求平方根的 sqrt( ) 函数等。这些常用函数都以静态方法的形式封装在 Math 类中。Math 类的使用比较简单。下面举一个简单的例子说明 Math 类的使用。

修改 Main 类的代码，在 main( ) 方法中计算一个幅度的 sin( ) 函数值、cos( ) 函数值等。修改后的 Main 类的代码如下（源代码为 08-05.java）：

```java
package org.example;
public class Main {
    public static void main(String[] args) {
        System.out.println(Math.sin(20));
        System.out.println(Math.cos(20));
        System.out.println(Math.sin(20)*Math.sin(20) + Math.cos(20)*Math.cos(20));
        System.out.println(Math.sqrt(20));
        System.out.println(Math.abs(-20));
    }
}
```

运行这个程序，显示以下结果。

```
0.9129452507276277
0.40808206181339196
1.0
4.47213595499958
20
```

8.2　字 符 串 类

字符串是编程实践中常用的数据类型之一。Java 中常用的字符串类型是 String 类和 StringBuffer 类。

8.2.1 String 类

Java 的 String 类对象是不可变的字符串对象,这里,"不可变"的含义是:一旦创建了一个 String 类的对象,则对象中的数据是无法变更的。创建 String 类对象的常用方法就是直接对一个 String 变量赋值:

```
String s1 = "Hello, World";
```

也可以使用传统方式:

```
String s2 = new String("Hello, World");
```

但是,Java 建议使用第一种方法。String 类的常用方法如表 8-2 所示。

表 8-2 String 类的常用方法

方　法	功　能
int indexOf(int ch) int indexOf(String s)	返回指定字符或字符串在字符串中的第一次出现索引位置,从 0 开始索引
int lastIndexOf(int ch) int lastIndexOf(String s)	返回指定字符或字符串在字符串中的最后一次出现索引位置,从 0 开始索引
int length( )	返回字符串的长度
boolean isEmpty( )	判断字符串是否是空串
String toLowerCase( )	将字符串的字符转换为小写字母
String[] split(String reg)	将字符串用指定正则表示进行分割,并将结果保存到新的字符串数组中
String subString(int start, int end)	返回按参数从字符串截取的一个子串
String trim( )	返回被截去字符串两端空格的一个新字符串对象
boolean equalsIgnoreCase(String s)	比较两个字符串是否相等,忽略大小写

关于字符串的使用举例,将在第 8.2.2 小节中介绍了 StringBuffer 类的使用后进行。

String 类和字符编码

8.2.2 StringBuffer 类

与 String 类不同,StringBuffer 类的对象是可变的,也就是说,可以改变 StringBuffer 对象的内容。StringBuffer 类类似于一个字符串容器,可以在 StringBuffer 对象中插入、删除、替换指定字符。StringBuffer 类的常用方法如表 8-3 所示。

表 8-3 StringBuffer 类的常用方法

方　法	功　能
StringBuffer append(String s) StringBuffer append(int i)	将字符串或指定数据作为字符串添加到 StringBuffer 字符串的末尾

续表

方　　法	功　　能
StringBuffer insert(int index, int i) StringBuffer insert(int index, String s)	在 StringBuffer 的指定位置插入新的字符串或将数据转换为字符串并插入指定位置
StringBuffer deleteCharAt(int index)	删除 StringBuffer 对象中指定位置的字符
StringBuffer delete(int start, int end)	删除 StringBuffer 对象指定的字符串子串
StringBuffer replace(int start, int end, String s)	用一个新的子串替换 StringBuufer 对象中的指定子串
StringBuffer reverse( )	反转 StringBuffer 对象中的字符串
String toString( )	将 StringBuffer 转换为 String 对象

现在举一个例子，介绍 String 类和 StringBuffer 类的使用。修改 Main 类，在其中的 main( ) 方法中创建 String 类和 StringBuffer 类的对象，对对象进行一些操作，然后显示相关信息。修改后的 Main 类的代码如下（源代码为 08-06.java）：

```java
package org.example;
public class Main {
    public static void main(String[] args) {
        String s1 = "Hello, World";
        String s2 = new String("HELLO, WORLD");
        System.out.println(s1.equalsIgnoreCase(s2));   // 显示 true
        System.out.println(s1.charAt(1));   // 显示字符 e
        System.out.println(s1.substring(1, 5));        // 显示字符串 ello
        System.out.println(s1);             // 显示 "Hello, World"
        System.out.println("----------------------");
        StringBuffer sb = new StringBuffer("Hello, World");
        sb.append(". 你好，中国");
        System.out.println(sb.toString()); // 显示 "Hello, World. 你好，中国"
        sb.replace(1, 2, "iii");
        System.out.println(sb.toString()); // 显示 "Hiiillo, World. 你好，中国"
        System.out.println("----------------------");
        String s3 = "one,two,three,four";
        String[] ss = s3.split(",");
        for(String s : ss) {
            System.out.println(s);
        }
    }
}
```

运行这个程序，显示以下信息。

```
true
e
ello
Hello, World
```

```
--------------------
Hello, World. 你好，中国
Hiiillo, World. 你好，中国
--------------------
one
two
three
four
```

8.3 随机数生成器类

获取随机数是很多程序都需要用到的功能，Java 提供了随机数生成器类。Random 类是 Java 经典的随机数生成器类，而 RandomGenerator 接口及其实现类则提供了随机效果更好的随机数。

8.3.1 Random 类

Random 类既可以生成随机整数，包括 int 类型和 long 类型的整数；也可以生成 float 类型和 double 类型的浮点数；还可以生成 boolean 类型的随机真假值。下面通过一个例子介绍 Random 类的使用。

可以使用两种方式创建 Random 类的对象：new Random( ) 和 new Random(long seed)，其中的参数 seed 指定 Random 类从 seed 参数指定的随机序列位置处开始返回随机数。修改后的 Main 类代码如下所示（源代码为 08-07.java）：

```java
package org.example;
import java.util.Random;
public class Main {
    public static void main(String[] args) {
        Random random = new Random(System.currentTimeMillis());
        System.out.println(random.nextInt(100));      //生成0到100之间的整数
        System.out.println(random.nextInt());         //生成0到最大int整数之间的整数
        System.out.println(random.nextFloat(100));    //生成0到100之间的浮点数
        System.out.println(random.nextFloat());       //生成0到1之间的浮点数
        System.out.println(random.nextBoolean());     //生成true或者false
    }
}
```

运行这个程序，显示以下结果。

32

```
1329620621
5.872607
0.024651289
true
```

注意：由于数据都是随机生成的，因此，每次运行程序显示的结果可能不同。

8.3.2 使用 RandomGenerator 接口生成随机数

RandomGenerator 是 Java 提供的用于生成随机数的接口，这个接口可以根据不同的随机数生成算法生成随机性不同的随机数。目前，JDK 中提供的随机数算法有 L128X1024MixRandom、L128X128MixRandom、L128X256MixRandom、L32X64MixRandom、L64X1024MixRandom、L64X128StarStarRandom、SplittableRandom 等。使用以下语句创建 RandomGenerator 接口的对象。

```
RandomGenerator rg1 = RandomGenerator.of("生成器类的名称");
```

下面举例说明 RandomGenerator 接口的使用。修改 Main 类，分别使用不同的随机数算法创建 RandomGenerator 接口对象，然后生成不同的随机数，修改后的 Main 类代码如下（源代码为 08-08.java）：

```java
package org.example;
import java.util.random.RandomGenerator;
public class Main {
    public static void main(String[] args) {
        RandomGenerator rg1 = RandomGenerator.of("L128X1024MixRandom");
        System.out.println(rg1.nextInt());       // 生成最小整数到最大整数之间的整数
        System.out.println(rg1.nextInt(20));     // 生成 0 到 20 之间的整数
        System.out.println(rg1.nextInt(40, 50)); // 生成 40 到 50 之间的整数
        System.out.println("--------------------------");
        RandomGenerator rg2 = RandomGenerator.of("L64X128StarStarRandom");
        System.out.println(rg2.nextFloat());       // 生成 0 到 1 之间的浮点数
        System.out.println(rg2.nextFloat(20));     // 生成 0 到 20 之间的浮点数
        System.out.println(rg2.nextFloat(40, 50)); // 生成 40 到 50 之间的浮点数
        System.out.println("--------------------------");
        RandomGenerator rg3 = RandomGenerator.of("SplittableRandom");
        System.out.println(rg3.nextDouble());       // 生成 0 到 1 之间的浮点数
        System.out.println(rg3.nextDouble(20));     // 生成 0 到 20 之间的浮点数
        System.out.println(rg3.nextDouble(40, 50)); // 生成 40 到 50 之间的浮点数
        System.out.println(rg3.nextBoolean());      // 随机生成 true 或者 false
    }
}
```

这段代码使用的随机数生成算法创建了 RandomGenerator 接口对象。运行这个程序，显示以下结果。

```
1425725054
11
43
---------------------------
0.40103102
6.757641
41.436962
---------------------------
0.9100438469192789
16.321796373349866
44.91894075375982
false
```

8.4 日期时间类

Java 的日期时间类获取日期时间对象，进行日期时间对象表示的日期差及进行日期时间格式的转换。

8.4.1 Date 类

既可以创建表示当前日期时间的 Date 对象，也可以根据指定的年、月、日、时、分、秒创建 Date 对象。Date 类的常用方法如表 8-4 所示。

表 8-4　Date 类的常用方法

方　　法	功　　能
Date( )	构造方法。根据当前日期时间创建 Date 的对象
Date(int y, int mon, int d,int h, int m, int sec)	构造方法。根据参数给定日期时间创建 Date 的对象
boolean after(Date when)	判断对象是否在参数对象所表示的时间之后
boolean befor(Date when)	判断对象是否在参数对象所表示的时间之前
long getTime( )	获取对象所表示的日期时间自 1970 年 1 月 1 日 0 时的毫秒数
void setTime(long millisecond)	将对象所表示的设置为参数所指定的时间，参数自 1970 年 1 月 1 日 0 时的毫秒数

在介绍完所有的日期时间类基本内容后，再举例说明 Date 类的使用。

8.4.2 Calendar 类

Calendar 类是表示日历的抽象类，不能使用 new 关键字创建 Calendar 类的对象，为此，

Calendar 类提供了一个创建 Calendar 实例的静态方法: getInstance( )。也就是说,可以使用以下语句创建 Calendar 对象的实例。

```
Calendar c = Calendar.getInstance();
```

使用 Calendar 对象,可以获取 Date 类的对象。

```
Date d = c.getTime();
```

Calendar 类提供了比 Date 类更多、更为灵活的日期时间操作方法。因此,在开发针对时间日期较为复杂的应用时,建议使用 Calendar 类。Calendar 类的常用方法如表 8-5 所示。

表 8-5 Calendar 类的常用方法

方法	功能
Date getTime( )	返回与 Calendar 对象表示的日期时间相同的 Date 对象
int get(int filed)	根据 field 的类型,返回 Calendar 对象中指定的日期时间数据。例如,当 filed 参数的值为 Calendar.YEAR 时将返回年份数据;当 field 参数的值为 Calendar.DAY_OF_MONTH 时将返回日期数据;当 filed 参数的值为 Calendar.DAY_OF_YEAR 时将返回对象所表示的日期是一年中的第几天等
boolean after(Object when)	判断对象是否在参数对象所表示的时间之后
boolean befor(Object when)	判断对象是否在参数对象所表示的时间之前
long getTimeInMillis( )	获取对象所表示的日期时间自 1970 年 1 月 1 日 0 时的毫秒数
void setTime(Date d)	将 Calendar 对象的日期时间设置为 Date 参数所表示的日期时间

在介绍完所有的日期时间类基本内容后,再举例说明 Calendar 类的使用。

8.4.3 SimpleDateFormat 类

因为各个国家和地区在表示日期时间时采用的格式不尽相同,所以,JDK 提供了专门进行日期格式转换的 SimpleDateFormat 类。在创建 SimpleDateFormat 类的对象时需要一个称为 pattern 的字符串参数来说明日期时间的格式。典型的创建 SimpleDateFormat 类的语句如下:

```
SimpleDateFormat sdf = new SimpleDateFormat("yyyy-MM-dd HH:mm:ss");
```

然后,可以使用 SimpleDateFormat 类的 format( ) 方法将给定的日期转换为指定的格式。例如,下列语句将显示日期时间信息。

```
Date d = new Date();
System.out.println(d);
System.out.println(sdf.format(d));
```

如果运行这几条语句，将显示以下结果。

```
Thu Jan 12 09:42:08 CST 2023
2023-01-12 09:42:08
```

同样是显示日期的信息，当使用了 SimpleDateFormat 类的对象将日期按指定的格式进行转换后。更适合国人的习惯。SimpleDateFormat 类的常用方法如表 8-6 所示。

表 8-6 SimpleDateFormat 类的常用方法

方　　法	功　　能
SimpleDateFormat(String pattern)	构造函数。用指定的格式串创建 SimpleDateFormat 类的对象。格式串的含义如下：yyyy 表示年，MM 表示月，dd 表示日，hh 表示 12 小时制，HH 表示 24 小时制，mm 表示分，ss 表示秒，S 表示毫秒
String format(Date d)	将日期 Date 参数表示的日期时间按指定的格式转换为字符串
Date parse(String date)	将用字符串表示的日期时间转换为 Date 对象

下面举个例子说明 Date 类、Calendar 类和 SimpleDateFormat 类的使用。修改 Main 类的代码，创建 Date 类的对象、Calendar 类和 SimpleDateFormat 类，并显示相关信息。修改后的 Main 类代码如下（源代码为 08-09.java）：

```java
package org.example;
import java.text.ParseException;
import java.text.SimpleDateFormat;
import java.util.Calendar;
import java.util.Date;
public class Main {
    public static void main(String[] args) throws ParseException {
        SimpleDateFormat sdf = new SimpleDateFormat("yyyy-MM-dd HH:mm:ss");
        Date d = new Date();
        System.out.println(d);
        System.out.println(sdf.format(d));
        System.out.println("--------------------------------");
        Date d2 = sdf.parse("2022-12-12 09:42:08");
        System.out.println(d2);
        System.out.println(sdf.format(d2));
        System.out.println("--------------------------------");
        Calendar c = Calendar.getInstance();
        c.setTime(d2);
        System.out.println("年:" + c.get(Calendar.YEAR));
        System.out.println("月:" + c.get(Calendar.MONTH));   //月份从0到11
        System.out.println("日:" + c.get(Calendar.DAY_OF_MONTH));
    }
}
```

运行这个程序，显示以下结果。

```
Thu Jan 12 09:58:50 CST 2023
2023-01-12 09:58:50
--------------------------------
Mon Dec 12 09:42:08 CST 2022
2022-12-12 09:42:08
--------------------------------
年:2022
月:11
日:12
```

8.5 使用 Java 基础类应用实践

JDK 文档对 Java 各种类的使用做了非常完整和非常权威的解释和说明，因此，要使用好 Java 的基础类和其他功能类，需要仔细阅读 JDK 文档。JDK 文档不仅对 Java 各种类的功能进行说明，还有一些使用举例可以参考，从而可以进一步提高对 Java 类的使用技巧。

8.6 案例：猜数游戏

猜数游戏是一个常见的小游戏：一个人的手里拿着一个数，请另一个人猜，允许对方猜 3 次，某一次如果猜对了就赢，否则提示是猜大了还是猜小了，如果 3 次都猜错了就输。编写一个 Java 程序模拟这个游戏过程。

8.6.1 案例任务

编写一个 Java 程序，先利用随机数生成器生成一个 100 以内的整数，然后要求用户从键盘接收一个 100 以内的整数，若两个数相等，则显示"成功"；否则，告诉用户所输入的数是大于还是小于所生成的随机数，直到用户猜成功为止。

8.6.2 任务分析

可以设计一个 Guesser 类，在该类的构造方法中使用随机数生成器生成一个 1 到 100 的随机数，然后，定义 guess 方法，提示玩家输入一个 1 到 100 的数并与预先生成的随机数进行比较，如果相等则玩家赢，否则玩家输。

8.6.3 任务实施

先创建 ch08-01 的 Java 工程，在其下创建名为 com.ttt.basic 的包。新建的 ch08-01 工

程如图 8-1 所示。

图 8-1 新建的 ch08-01 工程

在 com.ttt.basic 包下新建 Guesser 类，Guesser 类的代码如下（源代码为 08-10.java）：

```java
package com.ttt.basic;
import java.util.Scanner;
import java.util.random.RandomGenerator;
public class Guesser {
    private static final RandomGenerator rg = RandomGenerator.of("L128X1024MixRandom");
    private static final Scanner sc = new Scanner(System.in);
    private int secret;
    public Guesser() {
        secret = rg.nextInt(1, 101);
    }
    public boolean guess() {
        int count = 0;
        while(count < 3) {
            System.out.print("请猜数：");
            int n = sc.nextInt();
            if (n == secret)
                return true;
            else if (n > secret)
                System.out.println("猜大了");
            else
                System.out.println("猜小了");
            count++;
        }
        return false;
    }
}
```

修改 Main 类，在 main( ) 方法中创建 Guesser 对象并调用 guess( ) 方法开始游戏。修改后的 Main 类代码如下（源代码为 08-11.java）：

```
package org.example;
import com.ttt.basic.Guesser;
public class Main {
    public static void main(String[] args) {
        Guesser g = new Guesser();
        if (g.guess())
            System.out.println("你赢了");
        else
            System.out.println("你输了");
    }
}
```

运行这个程序，显示以下结果。

```
请猜数：50
猜小了
请猜数：75
猜大了
请猜数：60
猜小了
你输了
```

8.7 练习：计算闰年

编写一个 Java 程序，从键盘接收用户输入的日期数据，格式为 yyyy-mm-dd，然后，计算这个日期的年份是否是闰年。

第 9 章 Java 程序异常及程序调试技术

程序编码人员是难以避免编程过程中出现错误的，但是，应该采取必要的措施避免错误的发生，或者，即使出现了程序错误也能够快速找到错误并及时修正。本章对 Java 程序错误及其处理技术进行介绍。

9.1 程序错误分类

程序错误大致包括三类：其一，编译错误。这类错误也称为静态错误，一般是在编码过程中由于拼写原因或者未正确导入需要的 Java 程序类造成的。对于这类错误，IDEA 会及时告知编码人员，并且多数情况下都可以被快速修正。其二，程序异常。这类错误一般是由于对某些问题考虑不周导致的。对于这类错误，Java 专门提供类异常处理机制来处理这类错误，称为 Java 异常处理。其三，程序逻辑错误。这类错误一般是指程序的运行结果未与预期的结果一致。这类错误多数情况下是由于程序设计错误导致的，如解决问题的算法错误等。为了解决这类问题,需要重新审视程序算法,找到算法错误后再解决程序错误。

在第 1.4 节已经对如何解决程序第一类错误，也就是编译错误进行了说明。本章介绍如何处理第二类错误和第三类错误，也就是程序异常和程序运行逻辑错误的相关技术进行介绍。简单来说，可以使用 Java 异常处理技术处理程序异常；使用 Java 程序调试技术处理程序逻辑错误。

9.2 Java 程序异常及其处理入门

如前所述，Java 异常是指在编程过程中由于考虑不周而导致 Java 程序提前结束运行。在深入介绍 Java 异常及其处理技术之前，先通过一个简单例子了解一下 Java 异常。为此，在 IDEA 中新建一个名为 ch09-01 的工程，并在工程下新建名为 com.ttt.exception 的程序包。新建的 ch09-01 工程如图 9-1 所示。

9.2.1 Java 程序异常现象举例

这个例子非常简单：从键盘输入两个整数并相除，然后显示相除后的结果。为此，修改 Main 类，提示用户输入两个整数，并显示相除的结果。修改后的 Main 类如下（源代码为 09-01.java）：

图 9-1　新建的 ch09-01 工程

```
package org.example;
import java.util.Scanner;
public class Main {
    public static void main(String[] args) {
        Scanner sc = new Scanner(System.in);
        System.out.println("请输入一个整数：");
        int n1 = sc.nextInt();
        System.out.println("请再输入一个整数：");
        int n2 = sc.nextInt();
        int n3 = n1 / n2;
        System.out.println(n1 + " / " + n2 + " = " + n3);
    }
}
```

运行这个程序，输入的两个数分别为 100 和 50，运行结果如下：

```
请输入一个整数：
100
请再输入一个整数：
50
100 / 50 = 2
```

结果与预期一致。但是，如果输入的两个数分别为 100 和 0，运行结果如下：

```
请输入一个整数：
100
请再输入一个整数：
0
Exception in thread "main" java.lang.ArithmeticException: / by zero
    at org.example.Main.main(Main.java:14)
```

程序未能正常运行，而是显示以下异常信息。

```
Exception in thread "main" java.lang.ArithmeticException: / by zero
    at org.example.Main.main(Main.java:14)
```

这个异常信息表达了这样的含义：这个 Java 程序在运行过程中出现了称为算术异常的错误，这个错误的类型是 java.lang.ArithmeticException，原因是"/ by zero"被 0 除，在除法运算中，除数是不能为 0 的。

导致这个问题的原因是程序编码考虑不周：在程序中如果对输入第二个数进行检查，若为 0，则不进行两个数相除，而是告知输入数据错误，就不会出现这个错误。如果对代码做以下修改，则不会出现这个异常（源代码为 09-02.java）。

```
package org.example;
import java.util.Scanner;
public class Main {
    public static void main(String[] args) {
        Scanner sc = new Scanner(System.in);
        System.out.println("请输入一个整数：");
        int n1 = sc.nextInt();
        System.out.println("请再输入一个整数：");
        int n2 = sc.nextInt();
        if (n2 == 0) System.exit(1);    // 如果除数为 0，则退出程序
        int n3 = n1 / n2;
        System.out.println(n1 + " / " + n2 + " = " + n3);
    }
}
```

在这个代码中，增加了一条语句：

```
if (n2 == 0) System.exit(1);    // 如果除数为 0，则退出程序
```

检查除数是否为 0，若为 0，则直接退出程序。做这样的修改后，程序就不会再出现"被 0 除异常"。

对于某些异常可以采取如上所示的这种方法解决，也就是通过在代码中添加必要的检查代码来避免异常的发生。但是，对于某些异常是无法通过增加检查代码来避免的，例如，程序需要打开某个文件，而这个文件不小心被操作人员删除了。对于这种情况，Java 虚拟机在运行程序时，就会采用"抛出异常"的方式告知程序发生了某种异常状况，程序需要主动处理这种异常以避免程序被 Java 虚拟机提前终止运行。为此，Java 提供了专门对异常进行处理的机制，称为 Java 异常处理机制。

9.2.2 Java 异常处理入门

为了处理 Java 程序运行时异常，Java 提供了专门的机制，也就是 try...catch...finally 机制。例如，对于第 9.2.1 小节的程序，可以采用这种机制进行处理。修改后的代码如下（源代码为 09-03.java）：

```java
package org.example;
import java.util.Scanner;
public class Main {
    public static void main(String[] args) {
        Scanner sc = new Scanner(System.in);
        System.out.println("请输入一个整数:");
        int n1 = sc.nextInt();
        System.out.println("请再输入一个整数:");
        int n2 = sc.nextInt();
        try {
            int n3 = n1 / n2;
            System.out.println(n1 + " / " + n2 + " = " + n3);
        }
        catch(ArithmeticException e) {
            System.out.println("发生了程序异常,异常原因如下:" + e.getMessage());
        }
        finally {
            System.out.println("程序运行结束!");
        }
    }
}
```

Java 对异常进行处理的机制如下:

```
try {
    // 可能出现异常的代码
}
catch(异常类型1 变量) {
    // 对异常类型1进行异常处理的代码
}
catch(异常类型2 变量) {
    // 对异常类型2进行异常处理的代码
}
...
catch(异常类型n 变量) {
    // 对异常类型n进行异常处理的代码
}
finally {
    // 不管有无发生异常,都要被执行的代码
}
```

简单地说,Java 的异常处理机制就是:将可能出现异常的程序代码放置在 try 语句的大括号中,对可能出现的所有异常类型使用 catch 语句进行捕获并做适当处理,可以使用 finally 语句进行收尾清理处理,即不管程序是否出现异常,finally 代码块总是会被执行。

例如，在将输入的两个整数相除的例子中，使用以下代码。

```
try {
    int n3 = n1 / n2;
    System.out.println(n1 + " / " + n2 + " = " + n3);
}
catch(ArithmeticException e) {
    System.out.println("发生了程序异常，异常原因如下: " + e.getMessage());
}
finally {
    System.out.println("程序运行结束!");
}
```

将两个整数相除的语句放置在 try 代码块中，然后使用 catch 语句捕获 ArithmeticException 异常类型并进行适当处理。最后，不管有无发生异常，程序总是执行 finally 代码块，从而显示"程序运行结束!"的消息。运行这个程序，输入 100 和 10，程序运行结果如下：

```
请输入一个整数:
100
请再输入一个整数:
10
100 / 10 = 10
程序运行结束!
```

再次运行程序，输入 100 和 0，程序运行结果如下：

```
请输入一个整数:
100
请再输入一个整数:
0
发生了程序异常，异常原因如下: / by zero
程序运行结束!
```

了解了 Java 异常的基本概念和处理 Java 异常的基本方法后，为了更好地了解和处理 Java 异常，需要对 Java 的异常类体系有一个了解。

9.3 Java 程序异常及其处理进阶

Java 异常类构成了一个庞大的体系。异常类的顶端是 Throwable 类，典型异常类及其继承关系如图 9-2 所示。

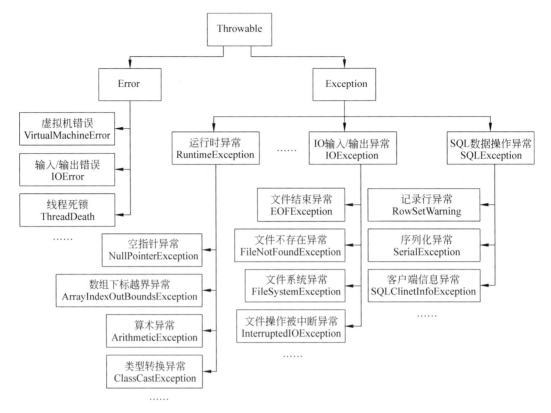

图 9-2 Java 典型异常类及其继承关系

下面对 Java 异常体系中的关键异常类进行介绍。

9.3.1 Error 类

JDK 文档对 Error 类是这样解释的：Error 类是 Trowable 的子类，它表示程序运行过程中出现了严重错误，并且应用程序不应该捕获它。

从这段描述可知：Error 是程序无法处理的错误，表示 Java 虚拟机在运行应用程序时发生了严重问题。大多数 Error 表示代码运行时 Java 虚拟机出现的问题，一般来说与代码编写者执行的操作无关，但是与程序对计算机资源的使用或占用有关。例如，当 Java 虚拟机不能为运行 Java 应用程序提供所需的内存资源时，将出现 OutOfMemoryError 错误。这些异常发生时，Java 虚拟机会终止程序运行。

解决 Java 程序运行时出现 Error 异常的有效途径就是仔细检查程序代码，减小程序对计算机资源的占用，或者补充缺失的资源。例如，使用 Java 程序进行图像处理时，由于图像占用过多计算机内存而导致出现 OutOfMemoryError 异常，能够解决这类问题的有效办法是缩小图像的尺寸或者采用新的算法来改善图像对内存的占用。再如，Java 虚拟机在运行某个 Java 程序时需要加载某个辅助程序类，可是这个辅助程序类由于某种原因不存在或者被删除了，Java 虚拟机就会抛出 NoClassDefFoundError 异常来告知异常发生点及其原因，能够解决这个异常的唯一办法就是提供需要的 Java 程序类。

9.3.2 Exception 类

Exception 类是 Java 程序可以捕获并需要适当处理的所有类的祖先类。常见的异常类，包括 RuntimeException 类、IOException 类、ArithmeticException 类、NullPoniterException 类、IndexOutOfBoundsException 类等都是 Exception 类的子类或孙类。根据程序中一个异常是否被 Java 编译器强制要求捕获并处理，可以进一步将 Exception 类及其子类划分为非检查性异常和检查性异常，习惯上称为 unchecked exception 和 checked exception。

检查性异常和非检查性异常的区分

9.3.3 非检查性异常

Java 编译器不强制程序捕获并处理的异常称为非检查性异常。非检查性异常有时也称为运行时异常。例如，虽然第 9.2.1 小节中例子的程序会产生 ArithmeticException 异常，但是编译器并不强制程序捕获并处理这个异常；程序可以自行选择是捕获并处理还是完全无视这个可能的异常。如果选择不捕获和处理非检查性异常，则当程序运行时出现了这类异常，程序会被终止运行。

非检查性异常一般都是 Error 类及其子类和 RuntimeException 类及其子类的子类。常见的非检查性异常包括 NullPointerException、IndexOutOfBoundsException、VirtualMachineError 等。图 9-3 所示是 JDK 文档中 RuntimeException 类的子类，这些类及其子类都是非检查性异常。

图 9-3　JDK 文档中列出 RuntimeException 类的子类

导致非检查性异常出现的原因，一般是编码的疏漏。因此，解决非检查性异常的办法是检查程序的疏漏，修正编码缺陷从而彻底解决它。

9.3.4 检查性异常

检查性异常也称为非运行时异常，是指编码过程中 Java 编译器会对程序出现的异常进行检查，如果程序存在检查性异常，那么编译器会提示编码人员，并且强制要求编码人员对这类异常进行处理,否则不能通过编译器编译,当然也就不能运行程序。

检查性异常包括 Exception 子类中除 RuntimeException 类及其子类以外的所有异常类,如 IOException 类及其子类、SQLException 类及其子类等。

下面通过一个例子来介绍什么是检查性异常,以及如何对检查性异常进行处理。

修改 Main 类,在 main( ) 方法中使用 Java 的 Class 类的静态方法加载一个程序类(不用关心这个类的作用,使用这个例子只是为了介绍检查性异常的相关概念),此时,IDEA 将出现如图 9-4 所示的提示,告知编码人员程序中有一个编译错误。

图 9-4 检查性异常提示

图 9-4 中两个箭头所指向的地方告知编码人员程序有编译错误。将光标放置到 forName( ) 方法上,将显示详细的错误原因,如图 9-5 所示。

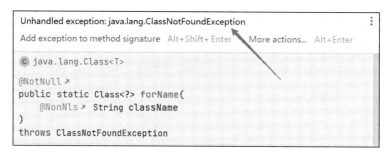

图 9-5 检查性异常的错误原因

选择图 9-5 中的 More actions → Surround with try/catch 命令,如图 9-6 所示。

图 9-6 使用 try...catch 捕获并处理异常

修改后的 Main 类代码如下（源代码为 09-04.java）：

```
package org.example;
public class Main {
    public static void main(String[] args) {
        try {
            Class.forName("com.ttt.exception.Abc");
        } catch (ClassNotFoundException e) {
            System.out.println("不能加载需要的类");
        }
    }
}
```

运行这个程序，由于所加载的类 com.ttt.exception.Abc 不存在，程序会抛出异常。运行结果如下：

不能加载需要的类

9.3.5 Java 异常处理

Java 异常处理机制需要注意以下事项。
（1）try 代码块中可以放置任何语句，包括可能产生异常的语句及不会产生异常的语句。
（2）catch 语句所捕获的异常必须"从小到大"。所谓"从小到大"，是指如果异常 A 是异常类 B 的子类，那么，必须将捕获 A 类异常的 catch 语句置于捕获 B 类异常的 catch 语句前面。
（3）在捕获异常的 catch 代码块中，可以对异常进行处理，也可以不经过任何处理直接通过 throw 语句将异常抛给方法的直接调用者。

9.3.6 自定义异常

如果 Java 提供的内置异常类不能满足程序设计的需求，可以设计自己的异常类。自定义异常类需要继承 Exception 类或其子类。
下面举一个例子说明如何自定义 Java 异常。在 ch09-01 工程的 com.ttt.exception 包下新建一个 MyException 的 Java 类，MyException 类的代码如下（源代码为 09-05.java）：

```
package com.ttt.exception;
public class MyException extends Exception {
    private ExceptionType type;
    public MyException(String message, ExceptionType type) {
        super(message);
        this.type = type;
```

```
    }
    public ExceptionType getType() {
        return type;
    }
}
```

在 MyException 类中，使用了枚举类型 ExceptionType 作为属性类型。ExceptionType 枚举类型的代码如下（源代码为 09-06.java）：

```
package com.ttt.exception;
public enum ExceptionType {
    Warning, Error, Fatal;
}
```

关于自定义异常的使用，将在第 9.4 节的案例程序中介绍。

9.4 案例：处理程序异常

本节通过一个综合性例子介绍 Java 异常的处理、自定义 Java 异常类的使用，同时，这个例子也使用了枚举类型。

9.4.1 案例任务

编写一个 Java 程序，定义一个 Worker 类，Worker 类包含以下属性：name、age、salary、company，分别表示姓名、年龄、每月薪金和所在公司名称。这个类只有使用 private 访问限定符修饰的构造方法。为了创建这个类的对象，需要定义使用 public static 限定符修饰的方法，进而可以创建 Worker 类的对象。这个方法接收从键盘输入的属性值后，通过私有的构造方法创建 Worker 对象。在私有构造方法中，对各个参数进行检查，并根据参数的合法性抛出不同类型的异常。要求：年龄为 18~60；薪金为 0~100000。编写一个测试程序，以测试 Worker 类的正确性。

9.4.2 任务分析

对案例任务要求进行分析可知，在 Worker 的私有构造方法中，需要对构造方法的参数进行检查，检查条件如下：

（1）年龄为 18~60。如果年龄不在这个合法区间内，则抛出 MyException 异常，其中的 type 类型设置为 ExceptionType.Age。

（2）薪金为 0~100000。如果薪金不在这个合法区间内，也抛出 MyException 异常，其中的 type 类型设置为 ExceptionType.Salary。

为 Worker 类编写一个 public、static 的方法，从键盘输入数据，并创建 Worker 类的对象，同时，再编写 toString 方法，将 Worker 对象转换为字符串。

9.4.3 任务实施

案例中用到的 ExceptionType 自定义异常类继续采用在第 9.3.6 小节中定义的 ExceptionType 异常类。在 ch09-01 工程的 com.ttt.exception 包下创建 Worker 类。Worker 类的代码如下（源代码为 09-07.java）：

```java
package com.ttt.exception;
import java.util.Scanner;
public class Worker {
    private String name;
    private int age;
    private float salary;
    private String company;
    private static Scanner sc = new Scanner(System.in);
    private Worker(String name, int age, float salary, String company)
    throws MyException {
        if ((age < 18) || (age > 60)) {
            throw new MyException("年龄错误", ExceptionType.Age);
        }
        if ((salary < 0) || (salary > 1000000)) {
            throw new MyException("薪金错误", ExceptionType.Salary);
        }
        this.name = name;
        this.age = age;
        this.salary = salary;
        this.company = company;
    }
    public static Worker getWorkerFromInput() throws MyException {
        System.out.print("请输入姓名：");
        String n = sc.nextLine();
        System.out.print("请输入年龄：");
        int a = sc.nextInt();
        System.out.print("请输入每月薪金：");
        float sa = sc.nextFloat();
        sc.nextLine();    //添加这条语句是为了清除上一条语句留下的换行符
        System.out.print("请输入公司名称：");
        String co = sc.nextLine();
        if ((n.isBlank()) || (co.isBlank()))
            return null;
        try {
            return new Worker(n, a, sa, co);
```

```java
        } catch (MyException e) {
            System.out.println("创建Worker对象失败!");
            System.out.print("失败原因:" + e.getMessage());
        }
        return null;
    }
    @Override
    public String toString() {
        return "Worker{" +
                "name='" + name + '\'' +
                ", age=" + age +
                ", salary=" + salary +
                ", company='" + company + '\'' +
                '}';
    }
}
```

在 Worker 类的私有构造方法中,当 age 参数或 salary 参数不符合合法性要求时,采用语句:

```java
if ((age < 18) || (age > 60)) {
    throw new MyException("年龄错误", ExceptionType.Age);
}
if ((salary < 0) || (salary > 1000000)) {
    throw new MyException("薪金错误", ExceptionType.Salary);
}
```

通过 throw 语句抛出 MyException 类型的异常。Java 规定,在代码中使用 throw 语句抛出异常时,要么使用 try...catch...finally 进行捕获处理,要么在方法中明确说明会抛出异常。这里采用第二种方式,即说明构造方法会抛出异常。

```java
private Worker(String name, int age, float salary, String company) throws MyException {
```

由于 getWorkerFromInput( ) 会调用 Worker 的构造函数创建 Worker 对象,同时,由于 Worker 的构造方法可能会抛出异常,因此,getWorkerFromInput( ) 方法需要对 MyException 异常进行处理:要么使用 try...catch...finally 进行捕获处理,要么在方法中明确说明会抛出异常。这里,采用 try...catch...finally 对异常进行处理。

```java
try {
    return new Worker(n, a, sa, co);
} catch (MyException e) {
    System.out.println("创建Worker对象失败!");
    System.out.print("失败原因:" + e.getMessage());
}
```

现在修改 Main 类，创建 Worker 类的对象，以便显示相关信息。修改后的 Main 类代码如下（源代码为 09-08.java）：

```java
package org.example;
import com.ttt.exception.Worker;
public class Main {
    public static void main(String[] args) {
        Worker w = Worker.getWorkerFromInput();
        System.out.println(w.toString());
    }
}
```

运行这个程序，输入 Gate、34、6000、TTT 后，运行结果如下：

```
请输入姓名：Gates
请输入年龄：34
请输入每月薪金：6000
请输入公司名称：TTT
Worker{name='Gates', age=34, salary=6000.0, company='TTT'}
```

再次运行这个程序，输入 Gates、34、6000，如果在程序提示输入公司名称时直接按 Enter 键，程序将显示以下结果。

```
请输入姓名：Gates
请输入年龄：34
请输入每月薪金：6000
请输入公司名称：
Exception in thread "main" java.lang.NullPointerException: Cannot invoke
 "com.ttt.exception.Worker.toString()" because "w" is null
    at org.example.Main.main(Main.java:8)
```

程序并没有按照预期显示 Worker 对象的信息，这是由于程序运行过程中出现了异常。通过对异常进行分析发现，在 Main 类的第 8 行出现了 NullPointerException 异常。为什么呢？其一，getWorkerFromInput( ) 方法在输入的名字或公司名称为空串时会返回 null 空对象，因此，在 Main 类的第 8 行得到的对象可能是 null 空对象；其二，Main 类并没有对非检查性异常进行捕获和处理。

解决 NullPointerException 异常的方法是避免出现空指针异常：对可能出现空指针的变量进行检查并做适当处理。修改后的 Main 类代码如下（源代码为 09-09.java）：

```java
package org.example;
import com.ttt.exception.Worker;
public class Main {
```

```java
    public static void main(String[] args) {
        Worker w = Worker.getWorkerFromInput();
        if (w == null) {
            System.out.println("不能创建 Worker 对象");
            return;
        }
        System.out.println(w.toString());
    }
}
```

再次运行程序，输入 Gates、34、6000，但是在程序提示输入公司名称时直接按 Enter 键，程序将显示以下结果。

```
请输入姓名：Gates
请输入年龄：34
请输入每月薪金：6000
请输入公司名称：
不能创建 Worker 对象
```

9.5 在 IDEA 中调试 Java 程序

所谓调试程序，简单来说就是一步一步地、一条语句一条语句地运行程序代码并观察程序的运行过程，其间观察程序变量值的变化，从而发现程序缺陷并修正它们。调试程序分为四步：第一步，设置程序断点；第二步，运行程序；第三步，检查程序变量并发现缺陷；第四步，修正缺陷并测试程序。如此循环。下面以第 9.4 节的案例程序为例介绍如何调试程序，并观察变量值的变化。

例如，在 Worker 类的第 38 行代码处设置程序断点。在 IDEA 代码编辑框的右边空白处单击，会出现一个红色的圆点，如图 9-7 所示。

图 9-7 设置程序断点

然后，单击 IDEA 工具栏中的调试按钮 开始调试程序，如图 9-8 所示。

图 9-8　调试程序

这时程序开始运行，会提示输入姓名、年龄、薪金和公司名称，然后在所设置的断点处自动停下来等待进一步操作，如图 9-9 所示。

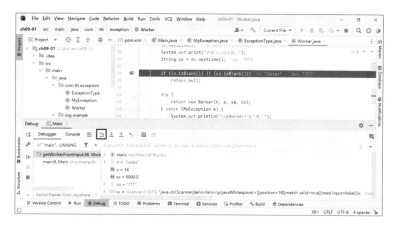

图 9-9　程序停留在断点处

此时，程序停留在黑底代码处，并在右下部窗口中显示各个变量的值。通过观察变量的实际值是否与预期值不同，可以发现程序缺陷。单击图 9-9 中的 按钮单步执行程序语句：单击一次这个按钮，就执行一条程序语句。执行一条语句后，显示如图 9-10 所示的界面。

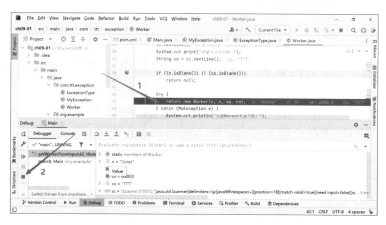

图 9-10　单步执行程序语句

再次观察程序变量的值。如此循环，观察变量的实际值与预期值的差异，找出程序缺陷并修正它们。如果通过调试发现了程序缺陷，则可以终止程序调试过程：在图9-10下方单击■按钮。

在 IDEA 中调试程序

9.6 Java 异常及程序调试应用实践

Java 异常包括 Error 类及其子类和 Exception 类及其子类。

对于 Error 类及其子类异常，虽然一般是由于 Java 虚拟机的原因，但是，本质上是由于 Java 程序的设计缺陷，特别是程序的算法设计缺陷导致的。对于这类异常，需要重新审视程序算法，减少对计算机资源，特别是内存资源的使用，从而从根本上杜绝这类问题。

Exception 类及其子类异常又可以划分为检查性异常和非检查性异常。对于检查性异常，由于 Java 编译器提供了较好的支撑，也就是 Java 编译器强制要求编码人员必须对这类异常进行妥善处理，因此，这类异常是相对比较容易处理的。对于非检查性异常，也就是运行时异常，处理起来相对困难：需要仔细审视程序代码，将每条语句运行时可能出现的各类非检查性异常都梳理出来，并从代码上对每个异常进行适当处理。这就是为什么在一个鲁棒的程序中，业务代码一般只占总代码量的30%~40%，而针对异常处理的代码却占到了60%~70%。

程序调试是发现并修正程序错误的重要技术手段。调试程序的方法非常简单但是非常有效。在具体调试一个程序之前，需要根据错误信息仔细分析程序结构，预估问题的可能位置，然后设置断点进行跟踪，进而发现并修正问题。

9.7 练习：将从键盘输入的字符串转换为浮点数

编写一个 Java 程序，该程序采用循环循环语句不断接收从键盘输入的任意字符串，将输入的字符串转换为浮点数并输出，输入 q 则结束。要求：需要处理程序运行时的各种异常。例如，如果输入的字符串是 "1234.5"，则程序可以将这个字符串转换为浮点数 1234.5 并显示出来；如果输入的字符串是 "X1234.5hello"，由于无法将这个字符串转换为浮点数从而产生异常，因此，需要程序捕获并适当处理程序出现的异常。

第 10 章 集合类及流式编程

集合类是 JDK 提供的非常有用的工具类。类似数组，集合类中可以存放多个元素。与数组不同，数组中的元素个数一旦确定是不可修改的，而集合类中可以存放的元素数目是动态的，可以随需要存放元素个数可变的数据。同时，结合 Java 集合类，出现了一种新的称为流式编程的编程模型。本章对 Java 的集合类及流式编程进行介绍。

10.1 泛　　型

Java 中的集合类都是泛型类，要合理灵活使用集合类，需要首先了解泛型。泛型允许程序员在编写代码时使用一些以后才指定的类型，在实例化时作为参数指明这些类型。这些解释理解起来有难度，下面通过一个例子来解释泛型及其应用。

10.1.1　泛型入门

考虑这样一种应用场景：定义一个类，这个类中有一个类型为数组的属性，要求在这个数组中可以存放任意类型的对象数据。例如，既可以存放字符串对象，也可以存放 Integer 对象，还可以存放自定义 Person 类的对象数据等。为了解决这类场景的应用需求，Java 提供了泛型：将类型作为参数的编程机制。

为了实现这个场景所需要的类，并展示程序运行效果，创建一个名为 ch10-01 的 Java 工程，并在这个工程下新建名为 com.ttt.generic 的包。新建的 ch10-01 工程如图 10-1 所示。

图 10-1　新建的 ch10-01 工程

在 com.ttt.generic 包下新建名为 MyArray 的 Java 类，在其中定义一个数组类型的属性，并编写必要的操作方法。MyArray 类的代码如下（源代码为 10-01.java）：

```java
package com.ttt.generic;
public class MyArray<E> {
    private E[] data;
    public MyArray(E... ds) {
        data = ds;
    }
    public void display() {
        for(E e:data) {
            System.out.println(e);
        }
    }
    public E get(int index) {
        if (index >= data.length)
            return null;
        else
            return data[index];
    }
}
```

由于要求定义的类能够存放任意类型的数组,这里的"任意类型"是在编码这个类时不能事先确定的,所以,在定义这个类时采用语句:

```java
public class MyArray<E> {
```

表示在 MyArray 这个类中有一个事先不能确定的类型,暂且用 E 来表示这个类型,因此,也称 E 为类型占位符。在类的定义中可以使用 E 来表示数组的类型。

```java
private E[] data;
```

在构造方法中可以使用 E 表示参数类型。

```java
public MyArray(E... ds) {
    data = ds;
}
```

在 display( ) 方法中使用 E 类型进行循环迭代。

```java
public void display() {
    for(E e:data) {
        System.out.println(e);
    }
}
```

已经定义了可以存放任何数据对象的 MyArray 类，现在测试这个类的正确性。为此，修改 Main 类的代码，创建 MyArray 类的对象并显示相关信息。修改后的 Main 类代码如下（源代码为 10-02.java）：

```java
package org.example;
import com.ttt.generic.MyArray;
public class Main {
    public static void main(String[] args) {
        MyArray<String> ma1 = new MyArray<>("One", "Two", "Three");
        ma1.display();
        System.out.println("-----------------------------");
        MyArray<Integer> ma2 = new MyArray<>(100, 200, 300);
        ma2.display();
    }
}
```

在 main( ) 方法中，通过语句：

```java
MyArray<String> ma1 = new MyArray<>("One", "Two", "Three");
```

创建了能够存放 String 类型的 MyArray 类的对象：此时，MyArray 类定义中的 E 就是字符串类型 String。也就是说，data 属性中存放的是 String 类型的元素。然后，调用这个对象的 display( ) 方法显示相关信息。类似地，使用语句：

```java
MyArray<Integer> ma2 = new MyArray<>(100, 200, 300);
```

创建了能够存放 Integer 类型的 MyArray 类的对象：此时，MyArray 类定义中的 E 就是整数类型 Integer。也就是说，data 属性中存放的是 Integer 类型的元素。然后，调用这个对象的 display( ) 方法显示相关信息。运行这个程序，显示以下信息。

```
One
Two
Three
-----------------------------
100
200
300
```

10.1.2 泛型类

在第 10.1.1 小节中定义的 MyArray 类就是典型的泛型类：在定义类时，在类的名称后面采用一对尖括号 "<>" 来指称类定义中用到的类型。尖括号中的类型占位符可以有多个，多个类型占位符之间用逗号","分隔。一旦在类定义中指称了类型占位符，则可以在类的属性、方法定义中将这些占位符作为普通类型使用。定义泛型类的一般语法如下：

```
访问限定符 class 类名<泛型标识，泛型标识，...> {
    // 非泛型属性定义："访问限定符 类型 属性名；"
    // 泛型属性定义："访问限定符 泛型标识 属性名；"
    // 构造方法
    // 其他方法
    ...
}
```

泛型中的类型参数名字用单个的大写字母来代替，常见类型参数名字：T，表示任意类；E，表示成员类型；K，与 V 搭配使用，表示键；V，与 K 搭配使用，表示值。例如，MyArray 泛型类使用 E 作为类型参数名来表述数组元素的类型。特别强调：Java 规定，泛型的类型参数，只能被传入类类型，不可以被传入基本数据类型。例如，在使用 MyArray 泛型类时，其中的参数 E 只能被传入 Java 的类类型，如 String、Integer 等，而不能是 int、long 等基本数据类型。

10.1.3 泛型方法

泛型可以应用到类定义的方法中。所谓泛型方法，是指在定义类的方法时可以为方法使用参数类型。定义泛型方法的一般语法如下：

```
访问限定符 <T，E，... > 返回值类型 方法名（形参列表）{
    // 方法体
}
```

从泛型方法的一般形式可以看出："访问限定符"与"返回值类型"中间的泛型标识符 <T，E，... > 是泛型方法的标志，只有这种格式声明的方法才是泛型方法；泛型方法声明时的泛型标识符 <T，E，... > 表示在方法中可以使用声明的泛型类型；泛型方法既可以定义在泛型类中，也可以定义在非泛型类中。

下面举个例子说明泛型方法的定义和使用。为此，在 ch10-01 工程的 com.ttt.generic 包中新建一个名为 MyGeneric 的类，在这个类中定义了一个泛型方法。MyGeneric 类的代码如下（源代码为 10-03.java）：

```
package com.ttt.generic;
public class MyGeneric {
    private String where;
    public MyGeneric(String where) {
        this.where = where;
    }
```

```
    public String getWhere() {
        return where;
    }
    public void setWhere(String where) {
        this.where = where;
    }
    public <T> void info(T t) {
        System.out.println(t.toString() + ":" + where);
    }
}
```

在 MyGeneric 类中定义了一个泛型方法。

```
public <T> void info(T t) {
    System.out.println(t.toString() + ":" + where);
}
```

用于显示 MyGeneric 对象的信息。现在修改 Main 类，创建 MyGeneric 类的对象，并调用 info( ) 方法显示信息。修改后的 Main 类代码如下（源代码为 10-04.java）：

```
package org.example;
import com.ttt.generic.MyGeneric;
public class Main {
    public static void main(String[] args) {
        MyGeneric gm = new MyGeneric(" 广州市 ");
        gm.info(" 你的通信地址 ");
    }
}
```

在调用 MyGeneric 对象的泛型方法时，由于传入的是字符串类型的参数，所以，此时 info( ) 方法的泛型类型 T 为 String 类型。当然，也可以传入其他参数类型给泛型 T。运行这个程序，显示以下信息。

```
你的通信地址 : 广州市
```

10.1.4　泛型接口

泛型也可以应用到接口定义中，从而为接口增加泛型类型参数。定义泛型接口的一般语法与定义普通接口非常相似，只是其中多了类型参数而已。定义泛型接口的一般形式如下：

```
访问限定符 inerface 接口名 < 泛型标识, 泛型标识, ...> {
    // 接口方法定义
```

```
        // 接口属性定义
}
```

在接口的方法定义和属性定义中均可以使用泛型参数。

有两种实现泛型接口的方式：其一，采用普通类实现泛型接口，此时，在实现类中需要明确给出接口的泛型类型参数；其二，采用泛型类实现泛型接口，此时，实现类的泛型类型要么与接口的泛型类型一致，要么包含接口的泛型类型。

下面举个例子说明泛型接口的定义和实现。为此，在 ch10-01 工程的 com.ttt.generic 包下新建一个名为 MyGenericInterface 的泛型接口，接口中只有一个方法。MyGenericInterface 的代码如下（源代码为 10-05.java）：

```
package com.ttt.generic;
public interface MyGenericInterface<T> {
    public abstract void info(T t);
}
```

首先使用普通类实现这个泛型接口。在 com.ttt.generic 包下新建名为 MyGenericInterfaceImpl2 的普通类，并实现 MyGenericInterface。MyGenericInterfaceImpl2 类的代码如下（源代码为 10-06.java）：

```
package com.ttt.generic;
public class MyGenericInterfaceImpl2 implements MyGenericInterface<String> {
    private float salary;
    public MyGenericInterfaceImpl2(float salary) {
        this.salary = salary;
    }
    @Override
    public void info(String s) {
        System.out.println(s + ": " + salary);
    }
}
```

在这个类的定义中，使用语句：

```
public class MyGenericInterfaceImpl2 implements MyGenericInterface<String> {
```

表示 MyGenericInterface 接口中的泛型类型 T 将被 String 类型所取代，因此，在类对接口 info( ) 方法实现的代码中：

```
@Override
```

```
public void info(String s) {
    System.out.println(s + ": " + salary);
}
```

方法中的 T 被 String 取代。

现在再使用一个泛型类来实现 MyGenericInterface 泛型接口。为此，在 com.ttt.generic 包下新建名为 MyGenericInterfaceImpl3 的泛型类。MyGenericInterfaceImpl3 代码如下（源代码为 10-07.java）：

```
package com.ttt.generic;
public class MyGenericInterfaceImpl3<T, E> implements MyGenericInterface<T> {
    private E e;
    public MyGenericInterfaceImpl3(E e) {
        this.e = e;
    }
    @Override
    public void info(T t) {
        System.out.println(t.toString() + ": " + e.toString());
    }
}
```

在这个类的定义中，使用语句：

```
public class MyGenericInterfaceImpl3<T, E> implements MyGenericInterface<T> {
```

表示实现类 MyGenericInterfaceImpl3 是个泛型类，其中"<T, E>"中的 T 必须与接口中的 T 一致，E 表示实现类 MyGenericInterfaceImpl3 有自己的一个泛型类型参数。因为 MyGenericInterfaceImpl3 是接口 MyGenericInterface 的泛型实现类，因此，在 info( ) 方法中仍然保留了泛型类型 T。

现在修改 Main 类，创建这两个实现类的对象，然后调用 info( ) 方法显示信息。修改后的 Main 类代码如下（源代码为 10-08.java）：

```
package org.example;
import com.ttt.generic.MyGenericInterfaceImpl2;
import com.ttt.generic.MyGenericInterfaceImpl3;
public class Main {
    public static void main(String[] args) {
        MyGenericInterfaceImpl2 mg2 = new MyGenericInterfaceImpl2(6000);
        mg2.info("我的薪金");
        MyGenericInterfaceImpl3<String, Integer> mg3 =
                                    new MyGenericInterfaceImpl3<>
                                    (5000);
```

```
            mg3.info(" 我今天行走的步数 ");
        }
}
```

运行这个程序，显示以下结果。

```
我的薪金：6000.0
我今天行走的步数：5000
```

10.1.5 泛型类型限制和泛型通配符 "?"

在定义泛型类、泛型方法及泛型接口时，除了可以使用 T、E、K、V 等这些符号指代无限制的参数类型外，还可以使用泛型类型限制机制以限制泛型参数类型。有两种方式，具体来说就是：其一，使用如 T、E、K、V 等这些符号指代无限制的参数类型；其二，使用 "T extends 父类" 指定泛型类型实参只能是指定父类或其子孙类。

在实例化泛型类型参数时，可以使用泛型通配符指代一定范围的参数类型。具体来说就是：其一，使用 "?" 匹配任意类；其二，使用 "? extends 父类" 匹配指定父类的所有父类及其子孙类；其三，使用 "? super 子类" 匹配指定类的父类及其祖先类。关于泛型通配符 "?" 的使用举例将在后续章节介绍。

10.2 集 合 类

Java 集合类是编程实践中常用的也是重要的内容。集合类，也称为容器类，在其中可以存放多个元素，并且存放于其中的元素个数是动态可变的。同时，由于 Java 集合类中定义的接口及实现类都采用了泛型，因此，集合类为保存数目可变并且类型可变的数据提供了便利。

观察 JDK 文档中的集合泛型类

10.2.1 集合类主要接口和类之间的关系

Java 的集合类本身是一个庞大的体系，其中包括可以存放单列值对象的 Collection 接口及其实现类和能够存放键值对对象的 Map 接口及其实现类。Java 集合类常用的接口和类及其之间的关系如图 10-2 所示。

Collection 接口的实现类用于存储单列数据对象。Collection 接口有两个子接口：List 接口和 Set 接口。其中，List 接口定义了元素内容可重复和有序的容器的操作方法，而 Set 接口则定义了元素内容不可重复和无序的容器的操作方法。常用的 List 接口的实现类有两个：其一为 ArrayList 类，使用动态数组实现了 List 接口；其二为 LinkedList 类，使用链表实现了 List 接口。常用的 Set 实现类也有两个：其一为 HashSet 类，使用对象的哈希值作为索引来存储对象；其二为 LinkedHashSet 类，使用对象的哈希值及链表来存储数据对象。

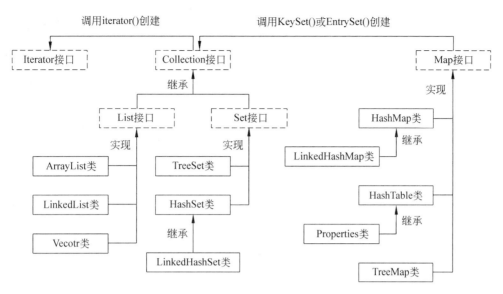

图 10-2　集合类常用的接口和类及其之间的关系

Map 接口的实现类用于保存名值对形式的数据。所谓名值对，就是为存储在 Map 实现类的每个数据元素都设置一个唯一的名字，通过这个名字可以快速地获取对应的数据对象。常用的 Map 接口的实现类有两个：HashMap 类及其子类 LinkedHashMap，其中，HashMap 采用哈希算法保存键值对的对应关系，而 LinkedHashMap 则采用哈希算法及链表方式保存键值对的对应关系。

10.2.2　List 接口及其实现类的使用

存储在 List 中的数据是有序和可重复的。这里有序的含义是：数据存储的顺序与数据访问的顺序是一致的。因此，对于 List 类型的容器，可以使用下标序号访问数据元素。常用的 List 接口的实现类是 ArrayList 类和 LinkedList 类。List 接口的常用方法如表 10-1 所示。

表 10-1　List 接口的常用方法

方　　法	功　　能
boolean add(E e)	将指定元素 e 添加到 List 的末尾，如果成功则返回 true，否则返回 false
void add(int index, E element)	将元素 element 插入由 index 索引指定的位置，List 中的 index 后的原有元素都后移一个位置
void clear( )	清除 List 中的所有数据元素
boolean contains(Object o)	检查 List 中是否包含对象 o，若包含则返回 true，否则返回 false
E get(int index)	返回 List 中指定索引位置的元素
int indexOf(Object o)	返回对象 o 在 List 中的索引，如果对象 o 不存在则返回 −1
Iterator<E> iterator( )	返回 List 对象的迭代器对象
static <E> List<E> of(E... elements)	创建一个包含给定元素，但是不可修改的 List 对象
E remove(int index)	从 List 中移除指定位置的元素，所有后续元素迁移一位

续表

方法	功能
int size( )	返回 List 中元素的个数
<T> T[] toArray(T[] a)	将 List 中的数据元素转换为泛型 T 类型的数组

下面举例说明 List 接口及其实现类的使用。为此，首先在 ch10-01 工程下新建一个名为 com.ttt.collection 的包，在这个包下新建一个名为 ListExample 的类，在这个类中定义 List 类型的属性，并在其中添加一些数据。ListExample 类的代码如下（源代码为 10-09.java）：

```java
package com.ttt.collection;
import java.util.ArrayList;
import java.util.Iterator;
import java.util.LinkedList;
import java.util.List;
public class ListExample {
    private List<String> list = null;
    public void constructByArrayListAndDisplay() {
        list = new ArrayList<>();
        list.add("Hello");
        list.add("World");
        list.add(1, "China");
        for (String s : list) {
            System.out.println(s);
        }
        System.out.println(list.contains("Hello"));
        list.remove(1);
        System.out.println(list.size());
    }
    public void constructByLinkedListAndDisplay() {
        list = new LinkedList<>();
        list.add("Hello");
        list.add("World");
        list.add(1, "China");
        list.remove(1);
        for(int i=0; i<list.size(); i++) {
            System.out.println(list.get(i));
        }
        Iterator<String> iter = list.iterator();
        while(iter.hasNext()) {
            System.out.println(iter.next());
        }
    }
}
```

在程序中，首先使用语句：

```
private List<String> list = null;
```

定义了 List 接口的变量，指定在 List 中可以存放 String 类型的数据元素，并初始化这个变量为 null。然后，在 constructByArrayListAndDisplay( ) 方法中，通过以下语句：

```
list = new ArrayList<>();
```

使 list 变量引用 ArrayList 类的对象。然后在 list 中添加一些 String 类型的元素，显示其中的元素信息，移除索引号为 1 的元素，最后显示 list 中的元素个数。类似地，在 constructByLinkedListAndDisplay( ) 方法中使用语句：

```
list = new LinkedList<>();
```

使 list 变量引用 LinkedList 类的对象。然后在其中添加一些元素，移除指定的元素，显示其中的元素信息。最后使用迭代器显示 list 中的元素。

```
Iterator<String> iter = list.iterator();
while(iter.hasNext()) {
    System.out.println(iter.next());
}
```

迭代器是一个 Iterator<E> 类型的对象，通过迭代器，可以顺序访问迭代器所指代的容器类型对象中的元素。迭代器的 hasNext( ) 方法用于判断迭代器所指代的对象中是否还有元素，而 next( ) 方法则用于获取下一个元素。

现在修改 Main 类，创建 ListExample 类的对象，并调用其中的方法以显示 list 中的信息。修改后的 Main 类代码如下（源代码为 10-10.java）：

```
package org.example;
import com.ttt.collection.ListExample;
public class Main {
    public static void main(String[] args) {
        ListExample le = new ListExample();
        le.constructByArrayListAndDisplay();
        System.out.println("--------------------");
        le.constructByLinkedListAndDisplay();
    }
}
```

运行这个程序，显示以下结果。

```
Hello
China
World
true
2
--------------------
Hello
World
Hello
World
```

10.2.3 Set 接口及其实现类的使用

存储在 Set 中的数据是无序和不可重复的。Java 的 Set 接口与数学中的 set 模型非常类似。Set 接口的方法与 List 接口非常类似,但是由于 Set 中的元素是无序的,因此不能使用下标操作 Set。常用的 Set 集合的实现类是 HashSet 类和 LinkedHashSet 类。Set 接口的常用方法如表 10-2 所示。

表 10-2 Set 接口的常用方法

方　　法	功　　能
boolean add(E e)	如果给定的参数 e 在 Set 中不存在,则将 e 添加到 Set 中。如果成功则返回 true,否则返回 false
void clear( )	清除 Set 中的所有数据元素
boolean contains(Object o)	检查 Set 中是否包含对象 o,若包含则返回 true,否则返回 false
Iterator<E> iterator( )	返回 Set 对象的迭代器对象
static <E> Set<E> of(E... elements)	创建一个包含给定元素,但是不可修改的 Set<E> 对象
E remove(Object o)	从 Set 中移除指定的对象 o
int size( )	返回 Set 中元素的个数
<T> T[] toArray(T[] a)	将 Set 中的数据元素转换为泛型 T 类型的数组

List 和 Set 可以包含任何类型的数据元素。在第 10.2.2 小节关于 List 的使用例子中,在 List 中存储了 String 类型的数据。为了演示 Set 接口及其实现类的使用,先定义几个具有继承关系的类: Animal 类、Cat 类、Corgi 类,分别表示动物类、狗类和威尔士柯基犬类。其中,Animal 类是 Cat 类的父类,Cat 类是 Corgi 类的父类。在 com.ttt.collection 包下新建 Animal 类、Cat 类、Corgi 类。Animal 类的代码如下(源代码为 10-11.java):

```java
package com.ttt.collection;
public class Animal {
    private String name;
    public Animal(String name) {
        this.name = name;
    }
```

```java
    public String getName() {
        return name;
    }

    public void setName(String name) {
        this.name = name;
    }

    @Override
    public String toString() {
        return "Animal{" +
                "name='" + name + '\'' +
                '}';
    }
}
```

Dog 类的代码如下（源代码为 10-12.java）：

```java
package com.ttt.collection;
public class Dog extends Animal {
    private String color;
    public Dog(String name, String color) {
        super(name);
        this.color = color;
    }
    public String getColor() {
        return color;
    }
    public void setColor(String color) {
        this.color = color;
    }
    @Override
    public String toString() {
        return "Cat{" +
                "name='" + getName() + '\'' +
                "color='" + color + '\'' +
                '}';
    }
}
```

Corgi 类的代码如下（源代码为 10-13.java）：

```java
package com.ttt.collection;
public class Corgi extends Dog {
    private int age;
```

```java
public Corgi(String name, String color, int age) {
    super(name, color);
    this.age = age;
}
public int getAge() {
    return age;
}
public void setAge(int age) {
    this.age = age;
}
@Override
public String toString() {
    return "Corgi{" +
            "name='" + getName() + '\'' +
            "color='" + getColor() + '\'' +
            "age=" + age +
            '}';
}
}
```

下面举例说明 Set 接口及其实现类的使用。为此，在 com.ttt.collection 包下新建一个名为 SetExample 的类，在这个类中定义 Set 类型的属性，并定义必要的方法。SetExample 类的代码如下（源代码为 10-14.java）：

```java
package com.ttt.collection;
import java.util.HashSet;
import java.util.LinkedHashSet;
import java.util.Set;
public class SetExample {
    private Set<Dog> set;
    public void createSetAndAddElement() {
        //set = new HashSet<>();
        set = new LinkedHashSet<>();
        // 编译错误，因为 set 中只能存放 Dog 及其子类对象
        //set.add(new Animal("大肥猫"));
        set.add(new Dog("大黄", "黄色"));
        set.add(new Dog("小黑", "黑色"));
        Corgi corgi = new Corgi("小黑", "黑色", 2);
        set.add(corgi);
        set.add(corgi); // 只能将第一个 corgi 对象加到 set 中，因为 set 中不允许出
                        //   现重复元素
    }
    public void display() {
        for(Animal e:set) {
```

```java
            if (e instanceof Corgi c ) {
                System.out.println(" 这是一只威尔士柯基犬 ");
                System.out.println(c.toString());
            }
            else if (e instanceof Dog d) {
                System.out.println(" 这是一只狗 ");
                System.out.println(d.toString());
            }
        }
    }
}
```

在 SetExample 代码中，使用语句：

```
private Set<Dog> set;
```

定义了可以存放 Dog 及其子类对象的 Set 属性，然后在 createSetAndAddElement( ) 方法中通过语句：

```
//set = new HashSet<>();
set = new LinkedHashSet<>();
```

创建了 Set 对象，可以使用 HashSet 类或者 LinkedHashSet 类创建 Set 对象。再通过语句：

```
// 编译错误，因为 set 中只能存放 Dog 及其子类对象
//set.add(new Animal(" 大肥猫 "));
set.add(new Dog(" 大黄 ", " 黄色 "));
set.add(new Dog(" 小黑 ", " 黑色 "));
Corgi corgi = new Corgi(" 韦基 ", " 黑色 ", 2);
set.add(corgi);
set.add(corgi); // 只能将第一个 corgi 对象加到 set 中，因为 set 中不允许出现重复元素
```

向 set 中添加一些元素。可是，以上语句出现了编译错误。

现在修改 Main 类，创建 SetExample 类的对象，调用相应方法添加元素，并显示信息。修改后的 Main 类代码如下（源代码为 10-15.java）：

```java
package org.example;
import com.ttt.collection.SetExample;
public class Main {
    public static void main(String[] args) {
        SetExample se = new SetExample();
        se.createSetAndAddElement();
```

```
        se.display();
    }
}
```

运行这个程序,显示以下结果。

```
这是一只狗
Cat{name='大黄', color='黄色'}
这是一只狗
Cat{name='小黑', color='黑色'}
这是一只威尔士柯基犬
Corgi{name='韦基', color='黑色', age=2}
```

10.2.4 Map 接口及其实现类的使用

存储在 Map 中的数据是键值对数据。所谓键值对,就是为存放在 Map 中的数据赋予一个名字,然后,可以通过每个数据的名字获取存入的对象数据值。常用 Map 集合的实现类是 HashMap 类和 LinkedHashMap 类。Map 接口的常用方法如表 10-3 所示。

表 10-3 Map 接口的常用方法

方　法	功　能
V put(K key, V value)	将参数给定的名值对 K/V 添加到 Map 中
V get(Object key)	如果 key 在 Map 中存在,则返回指定 key 对应的值对象;否则返回 null
V remove(Object key)	如果参数指定的 key 存在,则从 Map 中移除这个键值对数据
V replace(K key, V value)	将 Map 中 key 对应的值替换为 value 的值对象
void clear( )	清除 Map 中的所有元素
boolean containsKey(Object key)	判断 Map 中是否存在 key,如果存在则返回 true,否则返回 false
Set<K> keySet( )	以 Set 对象形式返回 Map 中所有的 key
int size( )	返回 Map 中的元素个数

下面举一个例子说明 Map 接口及其实现类的使用。在 com.ttt.collection 包下新建一个名为 MapExample 的类。MapExample 类的代码如下(源代码为 10-16.java):

```
package com.ttt.collection;
import java.util.HashMap;
import java.util.LinkedHashMap;
import java.util.Map;
import java.util.Set;
public class MapExample {
    static public void createMapAndDisplay() {
        Map<String, Animal> map = null;
        map = new HashMap<>();
```

```
        //map = new LinkedHashMap<>();
        map.put("0001", new Dog("小黑", "黑色"));
        map.put("0002", new Dog("小黄", "黄色"));
        map.put("0003", new Animal("小肥猫"));
        map.put("0004", new Corgi("小黑", "黑色", 3));
        Set<String> set = map.keySet();
        for(String key:set) {
            System.out.println(key + ": " + map.get(key).toString());
        }
    }
}
```

在 MapExample 类中的静态方法 createMapAndDisplay( ) 定义了一个 Map<String, Animal> 变量 map，map 的键 key 为 String 类型，value 值为 Animal 及其子类类型。然后使用语句：

```
map = new HashMap<>();
//map = new LinkedHashMap<>();
```

创建了 HashMap 对象，并向 map 中添加了一些信息。最后使用语句：

```
Set<String> set = map.keySet();
for(String key:set) {
    System.out.println(key + ": " + map.get(key).toString());
}
```

显示 map 中的信息：先创建 map 的 key 形成的 Set，再通过循环遍历 map，并显示每个 key 对应的 value 对象信息。

修改 Main 类，调用 MapExample 的静态方法，显示 map 的信息。修改后的 Main 类代码如下（源代码为 10-17.java）：

```
package org.example;
import com.ttt.collection.MapExample;
public class Main {
    public static void main(String[] args) {
        MapExample.createMapAndDisplay();
    }
}
```

运行这个程序，显示以下结果。

```
0004: Corgi{name='小黑', color='黑色', age=3}
0002: Cat{name='小黄', color='黄色'}
```

```
0003: Animal{name='小肥猫'}
0001: Cat{name='小黑', color='黑色'}
```

10.2.5 数组工具类 Arrays 的使用

Arrays 类提供了一系列用于对各类型数组进行操作的静态方法，如基于给定的数据创建 List 对象，对数组进行排序，从数组中查找指定的数据元素等。Arrays 类的常用静态方法如表 10-4 所示。

表 10-4 Arrays 类的常用静态方法

方　　法	功　　能
static <T> List<T> asList(T... a)	基于给定的参数创建 List 对象
static void sort(double[] a)	按照升序排序数组
static void sort(int [] a)	按照升序排序数组
static void sort(Object[] a)	对数组 a 按数组元素的自然顺序进行升序排序
static int binarySearch(int[] a, int key)	在数组 a 中查找指定的值 key，如果存在，则返回该值的索引，否则返回（−插入点 −1）值
static int binarySearch(float[] a, float key)	在数组 a 中查找指定的值 key，如果存在，则返回该值的索引，否则返回（−插入点 −1）值
static int binarySearch(double[] a, double key)	在数组 a 中查找指定的值 key，如果存在，则返回该值的索引，否则返回（−插入点 −1）值
static int binarySearch(Object[] a, Object key)	在数组 a 中查找指定的值 key，如果存在，则返回该值的索引，否则返回（−插入点 −1）值
static <T> T[] copyOf(T[] original, int newLength)	创建长度为 newLength 的数组，并将 original 数组的元素复制到新数组中。如果新数组的长度大于原数组，则用 null 填充
static boolean equals(long[] a, long[] a2)	比较两个数组的元素是否相等，如相等则返回 true，否则返回 false
static LongStream stream(long[] array)	将数组转换成一个 LongStream 对象
static <T> Stream<T> stream(T[] array)	将数组转换成一个 Stream<T> 对象
static String toString(Object[] a)	将数组转换为 String

下面修改 Main 类，在 main( ) 方法中创建一个 Dog 类的数组，然后基于 name 属性对数组排序，最后显示数组信息（源代码为 10-18.java）。

```
package org.example;
import com.ttt.collection.Dog;
import java.util.Arrays;
import java.util.Comparator;
public class Main {
    public static void main(String[] args) {
        Dog[] dogs = new Dog[]{
```

```
            new Dog("D1", "White"),
            new Dog("A2", "Black"),
            new Dog("C3", "Grey"),
            new Corgi("A2", "Yellow", 2)   //Corgi 是 Dog 的子类
    };
    //Arrays.sort(dogs, (o1, o2)->o1.getName().compareTo(o2.
    getName())); // 使用 lambda
    Arrays.sort(dogs, new Comparator<Dog>() {   // 使用匿名内部类
        @Override
        public int compare(Dog o1, Dog o2) {
            return o1.getName().compareTo(o2.getName());
        }
    });
    System.out.println(Arrays.toString(dogs));
    }
}
```

运行这个程序，显示以下信息。

```
[Cat{name='A2', color='Black'}, Corgi{name='A2', color='Yellow', age=2},
Cat{name='C3', color='Grey'}, Cat{name=' D1', color='White'}]
```

10.3 Java 流式编程

对集合对象的操作无非包括以下几种：向集合中添加新的元素、删除集合中的元素、获取集合中指定的元素、遍历集合中的元素。特别是为了遍历集合中的元素，需要不断地使用循环语句重复进行。循环是做事情的方式，而不是目的。为了更有效地实现集合的遍历，Java 提供了针对集合类的流式编程。

在流式编程中，大量使用了函数式接口和 Optional 类，因此，介绍流式编程之前，首先对 JDK 中常用的函数式接口和 Optional 类的使用进行介绍。

10.3.1 Java 常用函数式接口及其使用

如前所述，函数式接口是有且只有一个抽象方法的接口，通常，为了保证一个接口是函数式接口，Java 还提供了 @FunctionalInterface 注解。Lambda 表达式是实现函数式接口的有力工具。

函数式接口被大量应用于流式编程中，为此，JDK 提供了一系列常用的函数式接口。这些函数式接口包括 Supplier<T> 接口、Consumer<T> 接口、Predicate<T> 接口、Function<T,R> 接口、Runnable 接口、Comparator<T> 接口。除了 Runnable 接口外，其他接口函数式都是泛型接口，如表 10-5 所示。

表 10-5　JDK 的常用函数式接口

函数式接口（包+接口）	抽象方法原型	功　　能
java.util.function.Supplier<T>	T get( )	数据提供者：产生一个 T 类型的对象数据
java.util.function.Consumer<T>	void accept(T t)	数据消费者：处理一个 T 类型的数据，不返回任何值
java.util.function.Predicate<T>	boolean test(T t)	谓词：测试 T 类型的数据是否满足某个条件，是则返回 true，否则返回 false
java.util.function.Function<T,R>	R apply(T t)	功能函数：对 T 类型的参数进行处理并返回 R 类型的新对象数据
java.lang.Runnable	void run( )	执行特定操作：做特定的处理，不返回任何数据，这个接口常用于多线程程序中
java.util.Comparator<T>	int compare(T o1, T o2)	比较器：对两个 T 类型的数据做比较运算，返回整数类型的负数、0 或者正数，分别表示 o1 小于 o2、o1 等于 o2、o1 大于 o2

　　Lambda 表达式、接口方法引用和匿名内部类是实现函数式接口的常用方式，下面以 Supplier<T> 接口和 Consumer<T> 接口为例说明如何使用 Lambda 表达式、接口方法引用和匿名内部类实现函数式接口。为此，在 com.ttt.generic 包下新建一个名为 GeneralInterfaceExample 的类。GeneralInterfaceExample 类代码如下（源代码为 10-19.java）：

```
package com.ttt.collection;
import java.util.function.Consumer;
import java.util.function.Supplier;
public class GeneralInterfaceExample {
    private Supplier<Dog> supplier_lambda, supplier_anony, supplier_refer;
    private Consumer<Dog> consumer_lambda, consumer_anony, consumer_refer;
    public GeneralInterfaceExample() {
        supplier_lambda = () -> new Dog("大黄", "Yellow");
        supplier_anony = new Supplier<Dog>() {
            @Override
            public Dog get() {
                return new Dog("大黄", "Yellow");
            }
        };
        supplier_refer = GeneralInterfaceExample::getGog;
        consumer_lambda = (t) -> System.out.println(t);
        consumer_anony = new Consumer<Dog>() {
            @Override
            public void accept(Dog dog) {
                System.out.println(dog);
            }
        };
        consumer_refer = System.out::println;
    }
```

```
    public static Dog getGog() {
        return new Dog("大黄", "Yellow");
    }
    public void go() {
        Dog dog = supplier_lambda.get();
        consumer_refer.accept(dog);
        dog = supplier_anony.get();
        consumer_lambda.accept(dog);
        dog = supplier_refer.get();
        consumer_anony.accept(dog);
    }
}
```

在这个类中,首先使用语句:

```
private Supplier<Dog> supplier_lambda, supplier_anony, supplier_refer;
private Consumer<Dog> consumer_lambda, consumer_anony, consumer_refer;
```

定义了 Supplier<T> 接口和 Consumer<T> 接口变量:每个接口定义了三个变量,用于采用不同的方式实现该接口。然后,在构造方法中,分别使用三种方式对接口进行实现,使用语句:

```
supplier_lambda = () -> new Dog("大黄", "Yellow");
supplier_anony = new Supplier<Dog>() {
    @Override
    public Dog get() {
        return new Dog("大黄", "Yellow");
    }
};
supplier_refer = GeneralInterfaceExample::getGog;
```

采用三种方式对 Supplier<T> 接口进行实现并赋值给三个相应的变量。注意其中的语句:

```
supplier_refer = GeneralInterfaceExample::getGog;
```

采用了方法引用方式实现了接口。getDog( ) 方法是 GeneralInterfaceExample 类的一个静态方法。

```
public static Dog getGog() {
    return new Dog("大黄", "Yellow");
}
```

类似地,采用三种方式实现了 Consumer<T> 接口,并赋值给相应的变量。

```
consumer_lambda = (t) -> System.out.println(t);
consumer_anony = new Consumer<Dog>() {
    @Override
    public void accept(Dog dog) {
        System.out.println(dog);
    }
};
consumer_refer = System.out::println;
```

注意其中的语句:

```
consumer_refer = System.out::println;
```

采用了方法引用方式实现了接口。println( )方法是System类的静态成员out的一个实例方法。最后,在go( )方法中使用某一种方式产生Dog对象,再使用某一种方式处理所得到的对象数据。这里只是简单地将信息显示出来。

现在,修改Main类,创建GeneralInterfaceExample对象,然后调用其go( )方法。Main类的代码如下(源代码为10-20.java):

```
package org.example;
import com.ttt.collection.GeneralInterfaceExample;
public class Main {
    public static void main(String[] args) {
        GeneralInterfaceExample gie = new GeneralInterfaceExample();
        gie.go();
    }
}
```

运行这个程序,显示以下信息。

```
Cat{name='大黄', color='Yellow'}
Cat{name='大黄', color='Yellow'}
Cat{name='大黄', color='Yellow'}
```

10.3.2 Optional类及泛型通配符"?"使用举例

流式编程中大量使用了Optional类。Optional类是Java为了解决常见的NullPointerException空指针问题而提供的一个容器类。为了解释为什么Java会引入Optional类,看下面这个例子(源代码为10-21.java)。

```
package org.example;
```

```
import com.ttt.collection.Dog;
import java.util.Random;
public class Main {
    public static void main(String[] args) {
        Dog dog = getDogRandomly();
        System.out.println(dog.getName());
    }
    private static Dog getDogRandomly() {
        Random random = new Random();
        int chance = random.nextInt(4);
        if (chance%2 == 0)
            return new Dog("小金鱼", "蓝色");
        else
            return null;
    }
}
```

在这个例子中，方法 getDogRandomly( ) 根据随机数决定是返回一个 Dog 对象还是返回 null 空对象。而在 main( ) 方法中，使用语句：

```
System.out.println(dog.getName());
```

打印 animal 对象的 name 属性值。在某个时刻，当得到的 animal 是 null 对象时，程序运行就会产生 NullPointerException 空指针异常。当然，可以通过前面章节介绍的异常处理技术处理这个异常，但是，如前所示，更为推荐的处理方式是弥补程序漏洞。例如，可以使用下面的代码解决这个问题。

```
if (dog!= null)
    System.out.println(dog.getName());
```

在 Java 程序编程实践中，NullPointerException 空指针是非常普遍的现象。如果在每次得到一个对象时都做这样的检查，那么程序代码会显得非常烦琐。为此，Java 提供了 Optional 类解决这个问题。使用 Optional 重写上面这个例子的代码，修改后的代码如下（源代码为 10-22.java）：

```
package org.example;
import com.ttt.collection.Animal;
import com.ttt.collection.Dog;
import java.util.Optional;
import java.util.Random;
import java.util.function.Consumer;
public class Main {
    public static void main(String[] args) {
```

```
        Optional<Dog> od = getDogRandomly();
        od.ifPresentOrElse(System.out::println, ()->System.out.println
("对象为空"));
    }
    private static Optional<Dog> getDogRandomly() {
        Random random = new Random();
        int chance = random.nextInt(4);
        if (chance%2 == 0)
            return Optional.of(new Dog("小金鱼","蓝色"));
        else
            return Optional.empty();
    }
}
```

在这个程序中,getAnimalRandomly( )方法不再直接返回 Animal 对象,而是使用 Optional 类型将 Animal 包裹起来并返回 Optional<Animal> 对象。从而 getAnimalRandomly( ) 返回值永远不会为 null。此时,可以安全地使用 Optional 对象的方法获取 Animal 值,或者基于 Optional 对象进行其他操作。运行这个程序,当随机数为偶数时,显示以下信息。

```
Cat{name="小金鱼", color="蓝色"}
```

当随机数不为偶数时,显示以下信息。

```
对象为空
```

注意例子代码中这一条语句。

```
oa.ifPresentOrElse(System.out::println, ()->System.out.println("对象为空"));
```

这条语句完成对包裹在 Optional 对象中数据的显示。当然,这条语句比较难理解。要理解这条语句,需要对 Optional 类的方法有一定了解。Optional 类的常用方法如表 10-6 所示。

表 10-6 Optional 类的常用方法

方　　法	功　　能
static <T> Optional<T> of(T value)	基于 value 参数创建 Optional 对象
static <T> Optional<T> ofNullable(T value)	如果 value 不为 null,则创建包含 value 的 Optional 对象;否则创建包含空值的 Optional 对象
static <T> Optional<T> empty( )	创建包含空值的 Optional 对象
boolean isEmpty( )	如果 Optional 包含控制,则返回 true;否则返回 false
boolean isPresent( )	如果 Optional 中非空,则返回 true;否则返回 false
void ifPresent(Consumer<? super T> action)	如果 Optional 包含非空值,则执行 action 方法;否则不做任何操作

续表

方 法	功 能
void ifPresentOrElse(Consumer<? super T> action, Runnable emptyAction)	如果 Optional 包含非空值，则执行 action 方法；否则执行 emptyAction 方法
Optional<T> filter(Predicate<? super T>predicate)	如果 Optional 包含非空值，则执行 predicate 方法并返回新的 Optional 对象；否则直接返回包含空值的 Optional
T get( )	如果 Optional 中包含非空值，则返回这个值；否则抛出 NoSuchElementException 异常
<U> Optional<U> map(Function<? super T,? extends U> mapper)	如果 Optional 中包含非空值，则执行 mapper 操作并返回新的 Optional；否则直接返回包含空值的 Optional

注意表中的 ifPresentOrElse( ) 方法，该方法的原型如下：

```
void ifPresentOrElse(Consumer<? super T> action, Runnable emptyAction)
```

也就是说，ifPresentOrElse( ) 需要两个参数函数式接口参数：一个是 Consumer 接口参数，另一个是 Runnable 接口参数。在这个例子的语句：

```
oa.ifPresentOrElse(System.out::println, ()->System.out.println("对象为空"));
```

中，第一个参数使用了接口方法引用实参：使用 System 类的静态属性 out 的成员方法 println；第二个参数使用了 Lambda 表达式实参：使用 Lambda 表达式实现了 Runnable 接口。

作为泛型通配符的使用例子，再看看 Optional 类的 ifPresentOrElse( ) 方法。这个方法的第一个参数使用了泛型通配符 Consumer<? super T> action。这个泛型通配符要求，传递给 ifPresentOrElse( ) 方法的第一个实参必须是将 T 或者 T 的祖先类作为泛型类型实参的 Consumer 接口的实现类对象。对上面这个例子，传递给 ifPresentOrElse( ) 方法的第一个实参必须是将 Dog 的父类作为泛型类型实参的 Consumer 接口的实现类对象。为了便于理解，将上面这个例子修改为以下代码（源代码为 10-23.java）。

```
package org.example;
import com.ttt.collection.Animal;
import com.ttt.collection.Dog;
import java.util.Optional;
import java.util.Random;
import java.util.function.Consumer;
public class Main {
    public static void main(String[] args) {
        Optional<Dog> od = getDogRandomly();
        //od.ifPresentOrElse(System.out::println, ()->System.out.println("对象为空"));
        // 为了便于理解，可使用以下代码段取代上面这条语句
```

```
        Consumer<? super Dog> c = new Consumer<Animal>() {
            @Override
            public void accept(Animal animal) {
                System.out.println(animal);
            }
        };
        Runnable r = new Runnable() {
            @Override
            public void run() {
                System.out.println("对象为空");
            }
        };
        od.ifPresentOrElse(c, r);
    }
    private static Optional<Dog> getDogRandomly() {
        Random random = new Random();
        int chance = random.nextInt(4);
        if (chance%2 == 0)
            return Optional.of(new Dog("小金鱼", "蓝色"));
        else
            return Optional.empty();
    }
}
```

这段代码分别创建了 Consumer 接口对象和 Runnable 接口对象，然后将它们作为实参参数调用 oa 的 ifPresentOrElse( ) 方法。注意语句：

```
Consumer<? super Dog> c = new Consumer<Animal>() {
```

因为 Animal 类是 Dog 类的父类，所以是满足 "**? super Dog**" 这个泛型通配符要求的。如果将这条语句中的 Animal 改为 Corgi 类，则会产生编译错误：因为，Corgi 类不是 Dog 类的父类，因此，不满足泛型通配符 "**? super Dog**" 的规定。从这个例子也可以看出，使用接口方法引用和 Lambda 表达式可以极大地简化代码编写，并使得代码更加优雅。

"? super T" 和 "? extends T" 的含义和使用是类似的：只是此时，作为泛型实参的类型必须是 T 类型或其子孙类。

10.3.3 流式编程入门

先看一个简单的流式编程例子。这个例子创建以 Dog 对象为元素的 List 集合，然后将 List 中的 Dog 对象的颜色属性值全部改为大写，并从中查询是否存在给定名字的 Dog 对象。为此，在 ch10-01 工程中，修改 Main 类，在 Main 类的 main( ) 方法中实现这个功能。修改后的 Main 类代码如下（源代码为 10-24.java）：

```java
package org.example;
import com.ttt.collection.Corgi;
import com.ttt.collection.Dog;
import java.util.*;
public class Main {
    public static void main(String[] args) {
        List<Dog> list = new LinkedList<>();
        list.add(new Dog("D1", "White"));
        list.add(new Dog("A2", "Black"));
        list.add(new Dog("C3", "Grey"));
        list.add(new Corgi("A2", "Yellow", 2));
        Optional<Dog> d = list.stream()
                .peek((a) -> a.setColor(a.getColor().toUpperCase()))
                .filter((a)->a.getName().equalsIgnoreCase("A2"))
                .findFirst();
        if (d.isPresent())
           System.out.println(d.get());
        else
           System.out.println(" 不存在 ");
        //d.ifPresentOrElse(System.out::println, ()->System.out.
        println(" 不存在 "));
    }
}
```

这个程序先创建了一个 list 对象，再向其中添加一些元素对象。然后使用语句：

`list.stream()`

基于 list 对象创建 Stream 对象。这个 Stream 对象以 list 中的元素为数据，一旦创建了 Stream 对象，就可以在其上执行一系列操作。在本例中，在创建 Stream 对象之后，使用语句：

`.peek((a) -> a.setColor(a.getColor().toUpperCase()))`

对 Stream 中的每个数据采用 Lambda 表达式：

`(a) -> a.setColor(a.getColor().toUpperCase())`

将参数 a 对象的 color 属性转换为大写。之后，再使用语句：

`.filter((a)->a.getName().equalsIgnoreCase("A2"))`

针对 Stream 中的每个数据，采用 Lambda 表达式：

`(a)->a.getName().equalsIgnoreCase("A2")`

过滤出 name 属性等于 A2 的 Dog 对象并组成一个新的 Stream 对象。再通过语句：

```
.findFirst();
```

返回满足条件的第一个 Dog 对象并包裹在 Optional 对象中：

```
Optional<Dog> d
```

最后，使用语句：

```
if (d.isPresent())
    System.out.println(d.get());
else
    System.out.println(" 不存在 ");
```

显示检索到的 Dog 对象信息或者显示"不存在"的信息。更为简化地，可以将上面用于显示信息的这一大段语句修改为更为精简的代码。

```
d.ifPresentOrElse(System.out::println, ()->System.out.println(" 不存在 "));
```

运行这个程序，显示以下信息。

```
Cat{name='A2', color='BLACK'}
```

使用流式编程简化了代码结构，使代码更为简洁紧凑。例如，在这个例子中，本来需要两个循环完成的任务只需使用两个流的操作即可完成，因此，使用流式编程方式可以提高编程效率。

10.3.4　创建 Stream 和操作 Stream

在可以使用 Stream 进行流式编程之前，需要首先创建 Stream 对象；一旦创建了 Stream，就可以对 Stream 流施行一系列操作，包括 filter 过滤操作、map 映射操作、count 计数操作、limit 截取操作、forEach 注意迭代操作等。

创建 Stream 和操作 Stream

10.4　Java 数组、集合类及流式编程应用实践

数组和集合类为存储和处理大批量数据提供了便利。那么什么时候使用数组类型，什么时候使用集合类型呢？一般原则是：如果数据元素的数目一经创建不会发生变化，则建议使用数组类型，因为数组类型有较高的操作效率。但是，如果数据元素的数目在程序运

行过程中会发生变化，如插入新元素，或删除旧元素等，如果这种操作不频繁，则建议使用 ArrayList 类型的类来存储；反之，则应该采用 LinkedList 类型的类来存储。当然，如果元素不允许重复，则应该相应地使用 HashSet 类型或 LinkedHashSet 类型；如果需要存储和处理的数据是名值对形式的数据，则应该使用 Map 接口及其相应实现类。如果 Map 中的元素不会或者很少发生变化，则使用 HashMap 类；否则采用 LinkedHashMap 类。

流式编程是 Java 提高编程效率的一大举措，使用流式编程可以编写高效率的代码，但是，流式编程给代码的调试带来了难度。同时，由于流式编程大量地使用了 Java 的新机制，包括函数式接口、Lambda 表达式等，因此，需要编程者深入掌握相关知识和加以应用。记住，提高编程水平的最有效方式就是不断地进行实践、思考、总结。

10.5　案例：自制词典

本节通过自制词典案例程序总结集合类、流式编程的应用。词典是一款非常常用的工具软件，通过输入一个单词，词典给出单词的释义。

10.5.1　案例任务

使用 Java 的 Map 构建英文单词和中文释义之间的对应关系，然后，从键盘输入一个英文单词，查找对应的中文释义并输出。

10.5.2　任务分析

因为词典所存储的数据元素包括两列：英文单词和中文释义，因此，需要使用 Java 的 Map 接口及其实现类存储数据元素。同时考虑到一旦建立了英文单词和中文释义之间的关系，这个 Map 几乎就不会发生变化，因此，可以使用 Java 的 HashMap 类来存储这些数据元素。

10.5.3　任务实施

这个案例程序相对比较简单，直接在 Main 类中完成这些功能即可。为此，修改 Main 类代码，在 main( ) 方法中创建 HashMap 对象，并添加一些单词和其对应的中文释义的名值对到 HashMap 中。然后使用一个循环不断读取输入，并给出其对应的中文释义。修改后的 Main 类代码如下（源代码为 10-25.java）：

```java
package org.example;
import java.util.HashMap;
import java.util.Scanner;
public class Main {
    public static void main(String[] args) {
```

```java
HashMap<String, String> dictionary = new HashMap<>();
dictionary.put("hello", "你好。一种礼貌用语。");
dictionary.put("world", "世界。人类居住地的全称");
dictionary.put("water", "水。水是重要的资源");
dictionary.put("spring", "春天，弹簧。这个词有两种含义");
dictionary.put("snow", "雪。一种白色的东西");
dictionary.put("tree", "树木。一种高大的植物");
dictionary.put("city", "城市。车水马龙，人们集中居住的地方");
Scanner sc = new Scanner(System.in);
while(true) {
    String word, meaning;
    System.out.print("请输入英文单词：");
    word = sc.nextLine();
    if (word.isEmpty()) {
        System.out.println("再见!");
        break;
    }
    meaning = dictionary.get(word);
    if (meaning == null)
        System.out.println("对不起，词典里查不到这个单词：" + word);
    else
        System.out.println(word + ": " + meaning);
}
```

运行这个程序，显示以下结果。

```
请输入英文单词：spring
spring: 春天，弹簧。这个词有两种含义
请输入英文单词：water
water: 水。水是重要的资源
请输入英文单词：
再见！
```

10.6 练习：使用流式编程查询学生信息

定义一个 Student POJO 类，包含：姓名、年龄、家庭地址、电话号码等属性，然后，构建一个 List<E> 对象，包括多个学生对象。再编写一个测试类，从键盘输入学生姓名，要求使用流式编程从 List<E> 中查询出指定的学生对象并输出完整的学生信息。

第 11 章 文件输入 / 输出操作

需要程序处理的原始数据可以通过键盘输入，但是更多的情况是从文件中读取；类似地，程序运行的结果可以直接显示在屏幕上，但更多的情况是将程序运行的结果保存到文件中。这些操作都涉及文件的读写。本章对 Java 的文件输入 / 输出（I/O）操作进行介绍。

11.1 文件基本操作

在可以操作文件之前，一般都需要检查文件是否存在、文件中有多少信息内容等，Java 提供了 File 类、Files 类等完成这些操作。Files 类不仅可以操作文件属性，还可以方便地对文件内容进行读写。

11.1.1 使用 File 类操作文件属性

File 类是 Java 中用于操作目录或文件属性的最基本类。注意，File 类只能操作文件属性，不能进行文件内容读写。File 类的常用方法如表 11-1 所示。

表 11-1 File 类的常用方法

方　　法	功　　能
File(String pathname)	构造方法。基于给定的文件路径创建一个 File 对象
File(String parent, String child)	构造方法。基于给定的文件父路径和文件名创建一个 File 对象
boolean exists( )	判断文件或路径是否存在，如存在，则返回 true；否则返回 false
boolean createNewFile( )	创建文件
boolean mkdirs( )	如果路径中的目录不存在，创建路径中所有目录和子目录
boolean delete( )	删除文件或目录
String getName( )	获取文件名
boolean isDirectory( )	判断是否为目录，如是，则返回 true；否则返回 false
boolean isFile( )	判断是否为普通，如是，则返回 true；否则返回 false
long length( )	返回文件内容长度的字节数
String[] list( )	列表目录下的所有子目录和文件名
Path toPath( )	返回一个与 File 对象对等的 Path 接口对象

下面举例说明 File 类的使用。为此，在 IDEA 中新建一个名为 ch11-01 的 Java 工程。新建的 ch11-01 工程如图 11-1 所示。

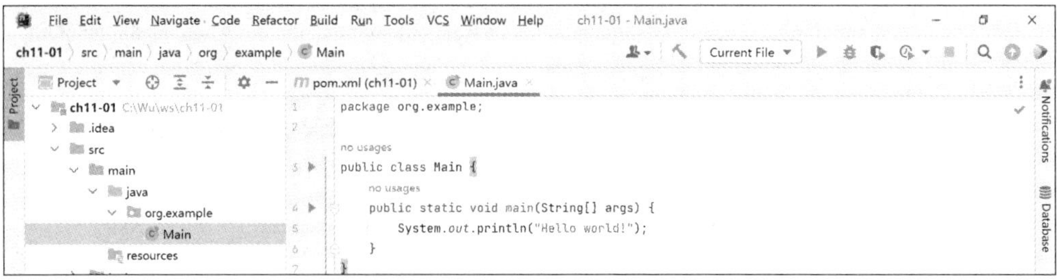

图 11-1 新建的 ch11-01 工程

修改 Main 类，在 main( ) 方法中创建 File 对象，然后基于这个 File 对象执行一系列操作。修改后的 Main 类代码如下（源代码为 11-01.java）：

```java
package org.example;
import java.io.File;
public class Main {
    public static void main(String[] args) {
        // 在 C:/Wu/Temp/ 目录下存在 data.txt 文件
        File file = new File("C:/Wu/Temp/data.txt");
        System.out.println(file.exists());      //true
        System.out.println(file.length());      // 显示文件长度
        System.out.println(file.getName());     //data.txt
        System.out.println(file.getParent());   //C:\Wu\Temp
        System.out.println("-------------------");
        File dir = new File("C:/Wu/Temp");      // 这是一个目录
        System.out.println(dir.isDirectory());// 显示 true
        System.out.println(dir.length());       // 基本目录块大小
        String[] ss = dir.list();               // 显示目录下的文件名和子目录名
        for(String s:ss)
            System.out.println(s);
    }
}
```

运行这个程序，显示以下结果。

```
true
63
data.txt
C:\Wu\Temp
-------------------
true
4096
data.txt
tools
```

11.1.2 使用 Files 类操作文件属性及读/写文件内容

File 类是较为经典的用于操作文件属性的类,但是 File 类不提供读写文件内容的操作。为此,Java 提供了一套新的称为 NIO(new input/output)的类,其中就包括 Files 类,通过这个类,不仅可以操作文件属性,也可以直接读写文件内容。

使用 Files 类操作文件属性及读/写文件内容

11.1.3 使用 WatchService 监视目录和文件变化

Java 的 NIO 提供了 WatchService 监控系统中文件的变化。它采用基于操作系统信号收发的监控机制,因此效率很高。WatchService 可以监视指定的 Path 对象(必须是目录对象)中文件的变化,包括创建新文件、删除文件、修改文件内容。

使用 WatchService 监视目录和文件变化

11.2 字节流读/写

虽然 Files 类提供了文件的读写操作,但是它提供的读写操作的控制粒度比较粗犷。例如,Files 类没有提供从文件中读取指定字节数数据的操作方法。为了实现对文件操作的精细粒度控制,Java 提供了一系列类和接口。

11.2.1 字节流的含义

本质上,存储于文件中的数据都是以字节形式存在的,这些字节数据一个字节接着一个字节地存储于文件中,就像流水一样连续存在,因此说文件是一个字节流。如果一个文件有 100 字节,那么这个文件就包含 100 字节的数据。使用 Java 的字节流操作类,可以非常准确地从文件中读取指定数目字节的数据。

字节流不仅可以存在于文件中,也可以存在于内存中。例如,如果使用以下语句定义了包含 100 字节的数组。

```
byte[] buffer = new byte[100]
```

也可以把这个字节数组看作一个字节流。

Java 提供了完整的操作字节流所需的类:处于字节流操作顶端的类是 InputStream 类和 OutputStream 类,它们都是抽象类。针对文件字节流输入/输出的最常用类是 FileInputStream 类和 FileOutputStream 类;针对内存字节流输入/输出的最常用类是 ByteArrayInputStream 类和 ByteArrayOutputStream 类。

11.2.2 读/写文件字节流

把文件看作字节流并进行操作的类包括 FileInputStream 和 FileOutputStream。其中，FileInputStream 用于从文件中读取字节数据，FileOutputStream 用于将字节数据写入文件中。因此，在创建 FileInputStream 和 FileOutputStream 对象时，均需要将一个文件名或 File 对象作为参数。FileInputStream 类和 FileOutputStream 类的常用方法分别如表 11-2 和表 11-3 所示。

表 11-2 FileInputStream 类的常用方法

方 法	功 能
FileInputStream(String name)	构造方法。通过指定文件的文件名创建流
FileInputStream(File file)	构造方法。通过指定的 File 对象创建流
int available( )	返回文件可读字节数
void close( )	关闭文件流
int read( )	从文件流中读 1 字节
int read(byte[] b)	从文件流中读取数组 b 的长度个数的字节并存储到 b 中，返回读取的字节数
int read(byte[] b, int off, int len)	从文件流中读取数组 len 字节并存储到 b 的从 off 开始的元素中，返回读取的字节数

表 11-3 FileOutputStream 类的常用方法

方 法	功 能
FileOutputStream(String name)	构造方法。通过指定文件的文件名创建流，若文件已经存在，则清除文件原有内容
FileOutputStream(String name, boolean append)	构造方法。通过指定文件的文件名创建流，在文件尾部追加新数据
void close( )	关闭文件流
void write(byte[] b)	将字节数组 b 的所有元素写入文件
void write(int b)	写一个字节到文件中
void write(byte[] b, int off, int len)	将数组 b 从 off 开始且长度为 len 的元素写入文件

下面举例说明 FileInputStream 类和 FileOutputStream 类的使用。这个例子先创建一个 FileOutputStream 类对象，然后向这个文件写入一个 byte 类型数组数据，最后关闭这个文件流。使用刚才这个文件再创建一个 FileInputStream 类的对象，从文件流中读取一个 byte 类型数组数据并显示。为此，修改 Main 类，再在 main( ) 方法中完成这些工作。修改后的 Main 类代码如下（源代码为 11-02.java）：

```
package org.example;
import java.io.FileInputStream;
import java.io.FileNotFoundException;
import java.io.FileOutputStream;
```

```java
import java.io.IOException;
public class Main {
    public static void main(String[] args) {
        FileOutputStream fos = null;
        try {
            fos = new FileOutputStream("C:/Wu/Temp/new.dat");
            byte[] b = new byte[10];
            for(int i=0; i<10; i++)
                b[i] = (byte)(100+i);
            fos.write(b);
        } catch (FileNotFoundException e) {
            System.out.println("程序运行出现了：FileNotFoundException 异常");
        } catch (IOException e) {
            System.out.println("程序运行出现了：IOexception 异常");
        }
        finally {
            if (fos != null) {
                try {
                    fos.close();
                } catch (IOexception e) {
                    System.out.println("程序运行出现了：RuntimeException异常");
                }
            }
        }
        FileInputStream fis = null;
        try {
            fis = new FileInputStream("C:/Wu/Temp/new.dat");
            byte[] b = new byte[10];
            int len = fis.read(b);
            System.out.println("读取了" + len + "字节");
            for(byte bt : b) {
                System.out.print(bt + " ");
            }
        } catch (FileNotFoundException e) {
            System.out.println("程序运行出现了：FileNotFoundException 异常");
        } catch (IOException e) {
            System.out.println("程序运行出现了：IOexception 异常");
        }
        finally {
            if (fis != null) {
                try {
                    fis.close();
                } catch (IOexception e) {
                    System.out.println("程序运行出现了：RuntimeException异常");
                }
```

```
            }
        }
    }
}
```

运行这个程序，显示以下结果。

```
读取了 10 字节
100 101 102 103 104 105 106 107 108 109
```

虽然这个程序可以正常运行，但是程序中用于异常处理的代码比正常处理数据的代码还要多。看看其中的一段：

```
FileOutputStream fos = null;
try {
    fos = new FileOutputStream("C:/Wu/Temp/new.dat");
    byte[] b = new byte[10];
    for(int i=0; i<10; i++)
        b[i] = (byte)(100+i);
    fos.write(b);
} catch (FileNotFoundException e) {
    System.out.println(" 程序运行出现了: FileNotFoundException 异常 ");
} catch (IOException e) {
    System.out.println(" 程序运行出现了: IOexception 异常 ");
}
finally {
    if (fos != null) {
        try {
            fos.close();
        } catch (IOException e) {
            System.out.println(" 程序运行出现了: RuntimeException 异常 ");
        }
    }
}
```

在这段代码中，因为创建 FileOutputStream 对象及进行流输出操作时均可能出现异常，因此，使用了两个用于捕获异常的 catch 语句。同时，因为完成流操作时需要关闭流，而在调用流的 close( ) 方法时也会出现异常，因此，又需要使用 try...catch 语句捕获异常，从而导致代码变得复杂和难以阅读。为此，Java 提供了 try-with-resource 机制来解决这个问题。

11.2.3 使用 try-with-resource 处理异常和关闭资源

try-with-resource 使用的一般格式如下：

```
try(创建资源语句；创建资源语句；...) {
    读写资源语句；
    ...
}
catch(异常捕获) {
    处理异常语句；
    ...
}
...
catch(异常捕获) {
    处理异常语句；
    ...
}
```

try-with-resource 机制与普通的异常捕获处理机制的不同点在于：try-with-resource 机制将创建资源的语句放在 try 语句的小括号中。通过这种处理后，Java 会自动关闭放置在 try 的小括号中打开的资源，而不需要编码者关闭。作为举例，现在使用 try-with-resource 机制修改第 11.2.2 小节的例子。修改后的代码如下（源代码为 11-03.java）：

```java
package org.example;
import java.io.FileInputStream;
import java.io.FileNotFoundException;
import java.io.FileOutputStream;
import java.io.IOException;
public class Main {
    public static void main(String[] args) {
        try(FileOutputStream fos = new FileOutputStream("C:/Wu/Temp/new.dat");) {
            byte[] b = new byte[10];
            for(int i=0; i<10; i++)
                b[i] = (byte)(100+i);
            fos.write(b);
        } catch (FileNotFoundException e) {
            System.out.println("程序运行出现了:FileNotFoundException异常");
        } catch (IOException e) {
            System.out.println("程序运行出现了:IOException异常");
        }
        try(FileInputStream fis = new FileInputStream("C:/Wu/Temp/new.dat");) {
            byte[] b = new byte[10];
            int len = fis.read(b);
            System.out.println("读取了" + len + "字节");
            for(byte bt : b) {
                System.out.print(bt + " ");
            }
```

```
        } catch (FileNotFoundException e) {
            System.out.println("程序运行出现了:FileNotFoundException 异常");
        } catch (IOException e) {
            System.out.println("程序运行出现了:IOException 异常");
        }
    }
}
```

修改后的代码比之前的代码简洁了不少。建议以后使用这种方式。

11.2.4 读/写内存字节流

将内存数据，如一个数组数据看作一个字节流并采用字节流形式进行读取和写入的常用 Java 类是 ByteArrayInputStream 类和 ByteArrayOutputStream 类。ByteArrayInputStream 类和 ByteArrayOutputStream 类的常用方法如表 11-4 和表 11-5 所示。

表 11-4 ByteArrayInputStream 类的常用方法

方　　法	功　　能
ByteArrayInputStream(byte[] buf)	构造函数。使用字节数组创建内存字节流
int available( )	返回字节流可用字节数
void close( )	关闭字节流
int read( )	读取 1 字节数据
int read(byte[] b, int off, int len)	从流中读取 len 字节并保存到 b 数组从 off 开始的位置，返回读取的字节数
byte[] readAllBytes( )	读取流中所有数据字节并返回读取结果

表 11-5 ByteArrayOutputStream 类的常用方法

方　　法	功　　能
ByteArrayOutputStream( )	构造方法。以默认长度的缓冲区创建一个新的内存字节流。当写入流的数据大于默认缓冲区长度时，缓冲区会自动增长
ByteArrayOutputStream(int size)	构造方法。创建一个新的长度为 size 的内存字节流。当写入流的数据大于 size 时，缓冲区会自动增长
void close( )	关闭字节流
void write(int b)	将字节 b 写入字节流
void writeBytes(byte[] b)	将数组 b 的全部元素写入字节流
void write(byte[] b, int off, int len)	将数组 b 从 off 开始且长度为 len 的数组元素写入流中
byte[] toByteArray( )	将流中所有数据转换成字节数组并返回

下面举例说明 ByteArrayInputStream 类和 ByteArrayOutputStream 类的使用。这个例子首先生成一些随机数，将这些随机数输出到一个 ByteArrayOutputStream 字节流中。然后，将 ByteArrayOutputStream 字节流中的数据转换为一个数组，在这个数组上创建一个 ByteArrayInputStream 字节流，再读取流中的数据并显示。为此，修改 Main 类，在 main( )

方法中完成这些功能。修改后的 Main 类代码如下（源代码为 11-04.java）：

```java
package org.example;
import java.io.ByteArrayInputStream;
import java.io.ByteArrayOutputStream;
import java.io.IOException;
import java.util.Random;
public class Main {
    public static void main(String[] args) {
        Random random = new Random(System.currentTimeMillis());
        byte[] b = null;
        try(ByteArrayOutputStream baos = new ByteArrayOutputStream();) {
            for(int i=0; i<20; i++) {
                baos.write(random.nextInt(100));
            }
            b = baos.toByteArray();
        }
        catch (IOException e) {
            System.out.println("程序异常，异常信息如下：");
            e.printStackTrace();
        }
        try(ByteArrayInputStream bais = new ByteArrayInputStream(b);) {
            while(bais.available() > 0) {
                System.out.print(bais.read() + " ");
            }
        }
        catch (IOException e) {
            System.out.println("程序异常，异常信息如下：");
            e.printStackTrace();
        }
    }
}
```

运行这个程序，显示以下信息。

```
18 96 57 27 45 6 63 25 34 69 54 15 63 6 1 88 24 91 40 7
```

11.3 字符流读／写

字符流，简单来说就是以可显示及可打印字符作为操作的基本元素。Java 提供了完整的操作字符流所需的类：处于字节流操作顶端的类是 Reader 类和 Writer 类，它们都是抽象类。对字符流的操作类大致分为两大块：不带缓冲的操作类和带缓冲的操作类。无缓冲的意思是，每次只能读取指定数目的字符，而不能按行方式来读取一行字符。常用的不带缓冲的操作类包括 FileReader 和 FileWriter；常用的带缓冲的操作类包括 BufferedReader 和 PrintWriter。

Java 默认以 Unicode 编码为基础处理字符数据。因此，在进行字符流操作之前，需要先对字符编码解码做简单介绍。

11.3.1 字符编码和字符解码

字符包括各种各样的文字，如英文字符、数字字符、中文字符等。为了在计算机中存储和处理字符数据，需要给每个字符一个唯一的编码。常见的字符编码包括 ASCII 码、ISO-8859 编码、Unicode 编码、UTF-8 编码、GBK 编码等。为了理解什么是字符编码，看看下面这个例子。为此，修改 Main 类为如下代码（源代码为 11-05.java）。

```java
package org.example;
import java.io.UnsupportedEncodingException;
public class Main {
    public static void main(String[] args) {
        // 默认情况下，Java 使用 Unicode 对字符进行编码，因此，本质上在计算机中保存
        的是这个字符的 Unicode 编码
        String s1 = "好";
        try {
            // 采用 GBK 对 "好" 编码
            byte[] bs_gbk = s1.getBytes("GBK");
            for (byte b : bs_gbk) {
                // 显示编码结果
                System.out.format("%x", b);
            }
            System.out.println();
            // 采用 UTF-8 对 "好" 编码
            byte[] bs_utf_8 = s1.getBytes("UTF-8");
            for (byte b : bs_utf_8) {
                // 显示编码结果
                System.out.format("%x", b);
            }
            System.out.println();
            // 对字节数组 bs_utf_8 采用与编码方式一致的方式解码，结果是正确的
            String s2 = new String(bs_utf_8, "UTF-8");
            System.out.println(s2);
            // 对字节数组 bs_utf_8 采用与编码方式不一致的方式解码，结果是错误的
            String s3 = new String(bs_utf_8, "GBK");
            System.out.println(s3);
        } catch (UnsupportedEncodingException e) {
            System.out.println("发生了异常，异常信息如下：");
            e.printStackTrace();
```

```
        }
    }
}
```

这个程序中，先定义了一个字符串变量并使之引用字符串"好"。

```
String s1 = "好";
```

之所以这里的字符串只有一个字符"好"，是因为希望观察一下"好"这个字符在各种不同编码方式下的编码结果。然后，采用以下语句。

```
// 采用 GBK 对"好"编码
byte[] bs_gbk = s1.getBytes("GBK");
for (byte b : bs_gbk) {
    // 显示编码结果
    System.out.format("%x", b);
}
```

对字符串"好"采用 GBK 方式进行编码并显示。接着，使用以下语句。

```
// 采用 UTF-8 对"好"编码
byte[] bs_utf_8 = s1.getBytes("UTF-8");
for (byte b : bs_utf_8) {
    // 显示编码结果
    System.out.format("%x", b);
}
```

对字符串"好"采用 UTF-8 方式进行编码并显示。然后，使用不同的方式对采用 UTF-8 对字符串"好"的编码结果进行解码。如下代码使用 UTF-8 进行解码。

```
// 对字节数组 bs_utf_8 采用与编码方式一致的方式解码，结果是正确的
String s2 = new String(bs_utf_8, "UTF-8");
System.out.println(s2);
```

而以下代码则使用与编码方式不一致的方式进行解码。

```
// 对字节数组 bs_utf_8 采用与编码方式不一致的方式解码，结果是错误的
String s3 = new String(bs_utf_8, "GBK");
System.out.println(s3);
```

运行这个程序，显示以下结果。

```
bac3
e5a5bd
好
濂�
```

从运行结果可以看出："好"的 GBK 编码结果是十六进制的 0xbac3，UTF-8 的编码结果是十六进制的 0xe5a5bd。针对"好"的 UTF-8 编码结果，如果采用 UTF-8 进行解码并显示，可以正常显示"好"，可是，如果采用 GBK 进行解码，则显示了乱码"濂�"。这说明一个重要规则：字符串的解码方式必须与其编码方式一致。

如前所述，默认情况下，Java 采用 Unicode 对字符串进行编码和解码。但是在某些特殊的场合，如从文件中读取一个字符串编码结果，需要使用与其一致的方式进行解码，否则解码结果是不正确的。了解了字符串编码和解码后，现在可以进行字符流的操作了。

11.3.2 无缓冲字符流读/写

无缓冲字符流操作最基础的类是 FileReader 类和 FileWriter 类，其中，FileReader 类用于从文件中读入字符，FileWriter 类则用于将字符写入文件中。FileReader 类和 FileWriter 类的常用方法如表 11-6 和表 11-7 所示。

表 11-6　FileReader 类的常用方法

方　　法	功　　能
FileReader(String fileName)	构造方法。基于给定的文件名和默认字符编码创建 FileReader 流
FileReader(String fileName, Charset charset)	构造方法。基于给定的文件名和给定字符编码创建 FileReader 流
void close( )	关闭文件字符流
String getEncoding( )	返回文件字符流的编码方式
int read( )	从文件流中读取一个字符，返回读取的字符个数
int read(char[] cbuf, int off, int len)	从文件流中读取 len 个字符并保存到 cbuf 中的从 off 开始处，返回读取的字符个数

表 11-7　FileWriter 类的常用方法

方　　法	功　　能
FileWriter(String fileName)	构造方法。基于给定的文件名和默认字符编码创建 FileWriter 流
FileWriter(String fileName, boolean append)	构造方法。基于给定的文件名和默认字符编码创建 FileWriter 流。以追加方式将字符写入流中
FileWriter(String fileName, Charset charset)	构造方法。基于给定的文件名和给定字符编码创建 FileWriter 流
FileWriter(String fileName, Charset charset, boolean append)	构造方法。基于给定的文件名和给定字符编码创建 FileWriter 流。以追加方式将字符写入流中
void close( )	关闭文件字符流
String getEncoding( )	返回文件字符流的编码方式

续表

方　　法	功　　能
void write(int c)	将字符 c 写入流中
void write(char[] cbuf, int off, int len)	将 cbuf 数组的从 off 开始且长度为 len 的字符写入流中
void write(String str, int off, int len)	将字符串 str 的从 off 开始且长度为 len 的字符写入流中

FileReader 和 FileWriter 的构造方法都可以使用一个 Charset 参数指定字符编码方式。Charset 提供了以下静态方法构建 Charset 对象。

```
static Charset forName(String charsetName)
```

其中，charsetName 的常用参数如表 11-8 所示。

表 11-8　charsetName 的常用参数

名　　称	描　　述
US-ASCII	ASCII 编码，不可编码中文字符
ISO-8859-1	ISO-8859-1 编码
UTF-8	UTF-8 编码
GBK	GBK 编码，中文字符编码方案
GB2312	GB2312 编码，简化版中文字符编码方案

下面举个例子说明 FileReader 类和 FileWriter 类的使用。这个例子先使用 FileWriter 将一些字符写入对应的文件中，然后基于这个文件创建一个 FileReader 的流，最后从流中读取字符并显示出来。为此，修改 Main 类，在 main( ) 方法中完成这些功能。修改后的 Main 类代码如下所示（源代码为 11-06.java）。

```java
package org.example;
import java.io.FileReader;
import java.io.FileWriter;
import java.io.IOException;
import java.nio.charset.Charset;
public class Main {
    public static void main(String[] args) {
        try(FileWriter fw = new FileWriter("C:/Wu/Temp/some.txt",
                Charset.forName("UTF-8"));) {
            fw.write('好');
            char[] chs = new char[]{'新','年','好','!'};
            fw.write(chs);
            fw.write("天天向上", 0, 4);
        } catch (IOException e) {
            System.out.println("程序运行异常。异常信息如下");
            e.printStackTrace();
```

```
        }
        try(FileReader fr = new FileReader("C:/Wu/Temp/some.txt",
                Charset.forName("UTF-8"));) {
            char[] chs = new char[100];
            int count = fr.read(chs);
            System.out.print("读取了" + count + "个字符:");
            for(int i=0; i<count; i++) {
                System.out.print(chs[i]);
            }
        } catch (IOException e) {
            System.out.println("程序运行异常。异常信息如下");
            e.printStackTrace();
        }
    }
}
```

运行这个程序，显示以下结果。

读取了 9 个字符: 好新年好！天天向上

11.3.3 带缓冲字符流读/写

常用的带缓冲的字符流操作类包括 BufferedReader 和 PrintWriter。其中，BufferedReader 类用于从字符流中读取字符，PrintWriter 则提供了直接以字符形式输出常见类型的数据到流中。BufferedReader 类和 PrintWriter 类的常用方法如表 11-9 和表 11-10 所示。

表 11-9 BufferedReader 类的常用方法

方 法	功 能
BufferedReader(Reader in)	构造方法。为指定的字符流 in 创建 BufferedReader 对象
void close( )	关闭流
Stream<String> lines( )	读取流中所有行，并以 Stream 方法返回
int read( )	从流中读取一个字符
int read(char[] cbuf, int off, int len)	从流中读取 len 个字符并存储到 cbuf 数组从 off 开始处的位置
String readLine( )	从流中读取一行数据并返回

表 11-10 PrintWriter 类的常用方法

方 法	功 能
PrintWriter(String fileName)	构造方法。使用指定的文件创建 PrintWriter 流
PrintWriter(String fileName, Charset charset)	构造方法。使用指定的文件和指定的编码方式创建 PrintWriter 流
void close( )	关闭流

续表

方　　法	功　　能
void print(char c)	输出一个字符到流中
void print(boolean b)	输出一个 boolean 值到流中
void print(double d)	输出一个 double 值到流中
void print(int i)	输出一个 int 值到流中
void print(String s)	输出一个字符串到流中
void println(long x)	输出一个 double 值到流中并换行
void println(char[] x)	输出一个字符数组到流中并换行
void println( )	输出一个换行符到流中

下面举一个例子说明 BufferedReader 类和 PrintWriter 类的应用。这个例子先以一个文件创建 PrintWriter 流，然后向这个流中输出一系列字符数据，然后，基于这个文件创建一个 BufferedReader 类对象，从流中读取字符数据并显示。为此，修改 Main 类，在 main( ) 方法中完成这些功能。修改后的 Main 类代码如下所示（源代码为 11-07.java）。

```java
package org.example;
import java.io.*;
import java.util.stream.Stream;
public class Main {
    public static void main(String[] args) {
        try(PrintWriter pw = new PrintWriter("C:/Wu/Temp/another.txt");) {
            pw.println(100);
            pw.println("Hello;");
            pw.println(" 你好 ");
            pw.println(1234.5f);
        } catch (FileNotFoundException e) {
            System.out.println(" 程序运行异常。异常信息如下 ");
            e.printStackTrace();
        }
        try(BufferedReader br = new BufferedReader(
                new FileReader("C:/Wu/Temp/another.txt"));) {
            Stream<String> all = br.lines();
            all.forEach(System.out::println);
        } catch (IOException e) {
            System.out.println(" 程序运行异常。异常信息如下 ");
            e.printStackTrace();
        }
    }
}
```

运行这个程序，显示以下信息。

```
100
Hello;
你好
1234.5
```

11.4 对象数据读/写

以对象为操作元素对流进行读写是 Java 提供的针对流的较高级操作方式。包括两个重要类：ObjectInputStream 类和 ObjectOutputStream 类。其中，ObjectInputStream 用于从流中读取对象数据；ObjectOutputStream 用于将对象数据写入流中。这两个类的常用方法如表 11-11 和表 11-12 所示。

表 11-11 ObjectInputStream 类的常用方法

方　　法	功　　能
ObjectInputStream(InputStream in)	构造方法。基于给定 in 参数创建对象
int available( )	返回流中可用数据量大小
void close( )	关闭流
byte readByte( )	从流中读取 1 字节
double readDouble( )	从流中读取一个 double 数据
int readInt( )	从流中读取一个 int 数据
long readLong( )	从流中读取一个 long 数据
final Object readObject( )	从流中读取一个对象

表 11-12 ObjectOutputStream 类的常用方法

方　　法	功　　能
ObjectOutputStream(OutputStream out)	构造方法。基于给定 out 参数创建对象
void close( )	关闭流
void write(byte[] buf)	将字节数组 buf 写入流中
void write(int val)	将一个字节写入流中
void writeDouble(double val)	将一个 double 数据写入流中
void writeInt(int val)	将一个 int 数据写入流中
void writeBytes(String str)	将一个字符串数据写入流中
final void writeObject(Object obj)	将一个对象序列化并写入流中

下面举例说明 ObjectInputStream 类和 ObjectOutputStream 类的使用。这个例子首先基于一个文件创建 ObjectOutputStream 类的对象，然后往流中写入一些数据，包括一个 int 整数、一个 float 数据，再写入一个 User 对象数据。最后，基于这个文件创建 ObjectOutputStream 类的对象，读取其中的数据并显示。为此，修改 Main 类为以下代码（源代码为 11-08.java）。

```java
package org.example;
import java.io.*;
public class Main {
    public static void main(String[] args) {
        try(FileOutputStream fos = new FileOutputStream("C:/Wu/Temp/object.dat");
            ObjectOutputStream oos = new ObjectOutputStream(fos)) {
            oos.writeInt(1000);
            oos.writeFloat(1234.5f);
            User u = new User("张三", 20);
            oos.writeObject(u);
        } catch (IOException e) {
            System.out.println(" 程序运行异常。异常信息如下 ");
            e.printStackTrace();
        }
        try(FileInputStream fis = new FileInputStream("C:/Wu/Temp/object.dat");
            ObjectInputStream ois = new ObjectInputStream(fis)) {
            int i = ois.readInt();
            float f = ois.readFloat();
            User u = (User)ois.readObject();
            System.out.println(i);
            System.out.println(f);
            System.out.println(u);
        } catch (IOException e) {
            System.out.println(" 程序运行异常。异常信息如下 ");
            e.printStackTrace();
        } catch (ClassNotFoundException e) {
            throw new RuntimeException(e);
        }
    }
    private static class User implements Serializable{
        private String name;
        private int age;
        public User(String name, int age) {
            this.name = name;
            this.age = age;
        }
        @Override
        public String toString() {
            return "User{" +
                    "name='" + name + '\'' +
                    ", age=" + age +
                    '}';
        }
    }
}
```

如果需要将类的对象作为流数据写入流中，需要这个类实现 Serializable 接口，因此在定义 User 类时需要指明这一点。

```
private static class User implements Serializable{
```

Serializable 接口是一个标识性接口，其中没有任何抽象方法。现在，运行这个程序，显示以下信息。

```
1000
1234.5
User{name='张三', age=20}
```

11.5 Java 流操作应用实践

不要将本章介绍的流与第 10.3 节介绍的流式操作混淆。本章介绍的流是指数据：数据就像流水一样，可以对数据进行读写操作；而第 10.3 节中介绍的流式操作则是指在数据对象上进行连续的方法调用以实现对数据对象进行连续操作。

Java 流是一个庞大的框架体系。其中包含了众多的针对字节流、字符流和对象流的操作类和接口。关于在工程实践中如何选择和使用，给出以下应用实践。

（1）如果只需读取文件的基本属性信息、判断文件或目录是否存在、读取目录下的所有文件名和子目录名等，则 File 类已经足够满足需求。

（2）如果只需要对字符流进行读写，那么，BufferedReader 和 PrintWriter 是较好的选择，因为，它们都提供了便利的读写字符数据的方法。

（3）如果需要对包括各种不同数据类型的复杂数据对象进行存储和读取，那么，只有 ObjectInputStream 和 ObjectOutputStream 类满足需求，因为它们提供了便利的读写任何类数据的方法。

（4）如果需要将从一个流中读取的字节数据保存到内容以便进行后续的操作，那么，ByteArrayInputStream 和 ByteArrayOutputStream 是最佳选择，因为，它们提供了可以将流中的字节数据以数组形式进行操作的便利方法。

11.6 案例：通讯录程序

通讯录程序是一款常见的信息管理工具，通过它可以完成对朋友、公司等的通讯录信息的管理。

11.6.1 案例任务

在通讯录中，一个人公司的基本信息包括名称、电话号码、通讯录地址等信息。设计一个简单的通讯录程序,可以实现通讯录信息的增加、删除、修改和查询。要求:将通讯录信息保存在文件中,并且,在启动程序时从文件中读取所有的通讯录信息,在退出程序时,将通讯录信息保存到文件中。

11.6.2 任务分析

首先需要定义 Contact 类用于存储一条通讯录信息。然后,由于需要将通讯录信息写入文件,考虑到需要操作复杂的对象信息流,因此,使用 ObjectInputStream 和 ObjectOutStream 类来操作对象信息流。最后,定义一个 Manager 类,将通讯录数据从流中读取到 List 中,并且完成增删改查相关操作。

11.6.3 任务实施

在 ch11-01 工程下新建一个名为 com.ttt.contact 的包,在其下新建 Contact 类用于存储通讯录信息,再新建一个 Manager 类用于完成对通讯录信息的增删改查。Contact 类的代码如下(源代码为 11-09.java):

```java
package com.ttt.contact;
import java.io.Serializable;
public class Contact implements Serializable {
    private String name;
    private String phone;
    private String addr;
    public Contact(String name, String phone, String addr) {
        this.name = name;
        this.phone = phone;
        this.addr = addr;
    }
    public String getName() {
        return name;
    }
    @Override
    public String toString() {
        return "Contact{" +
                "name='" + name + '\'' +
                ", phone='" + phone + '\'' +
                ", addr='" + addr + '\'' +
                '}';
    }
}
```

Manager 类的代码如下(源代码为 11-10.java):

```java
package com.ttt.contact;
import java.io.*;
import java.util.*;
import java.util.function.Consumer;
public class Manager {
    List<Contact> list;
    Scanner sc;
    public Manager() {
        list = new LinkedList<>();
        sc = new Scanner(System.in);
    }
    public void run() {
        try (FileInputStream fis = new FileInputStream("C:/Wu/Temp/contact.dat");
            ObjectInputStream ois = new ObjectInputStream(fis)) {
            while (true) {
                Contact c = (Contact) ois.readObject();
                list.add(c);
            }
        } catch (EOFException e) {
        } catch (IOException | ClassNotFoundException e) {
            System.out.println("目前没有任何联系信息");
        }
        int choice = -1;
        while (choice != 0) {
            choice = getActionChoice();
            switch (choice) {
                case 1 -> query();
                case 2 -> add();
                case 3 -> modify();
                case 4 -> delete();
                case 0 -> {
                    break;
                }
                default -> System.out.println("非法操作!");
            }
        }
        try (FileOutputStream fos = new FileOutputStream("C:/Wu/Temp/contact.dat");
            ObjectOutputStream oos = new ObjectOutputStream(fos)) {
            list.forEach(new Consumer<Contact>() {
                @Override
                public void accept(Contact contact) {
                    try {
                        oos.writeObject(contact);
```

```java
                } catch (IOException e) {
                    System.out.println("程序发生异常, 异常信息: ");
                    e.printStackTrace();
                    return;
                }
            }
        });
    } catch (IOException e) {
        System.out.println("程序发生异常, 异常信息: ");
        e.printStackTrace();
        return;
    }
    System.out.println("ByeBye!");
}
private int getActionChoice() {
    System.out.println("Menu");
    System.out.println("1: 查询通讯录信息 ");
    System.out.println("2: 增加通讯录信息 ");
    System.out.println("3: 修改通讯录信息 ");
    System.out.println("4: 删除通讯录信息 ");
    System.out.println("0: 退出 ");
    int choice = -1;
    while (choice < 0) {
        try {
            System.out.print(" 请输入你的选择: ");
            choice = sc.nextInt();
        }
        catch (InputMismatchException e) {}
        finally {
            sc.nextLine(); // 放弃输入整数后的回车符
        }
    }
    return choice;
}
private void query() {
    System.out.print(" 请输入名称: ");
    String name = sc.nextLine();
    Optional<Contact> found = list.stream().
            filter(c -> c.getName().equalsIgnoreCase(name)).findFirst();
    found.ifPresentOrElse(System.out::println,
        () -> System.out.println(" 没有这个联系人 "));
}
private void add() {
    System.out.print(" 请输入名称: ");
    String name = sc.nextLine();
```

```java
            System.out.print("请输入电话号码: ");
            String phone = sc.nextLine();
            System.out.print("请输入通讯录地址: ");
            String addr = sc.nextLine();
            list.add(new Contact(name, phone, addr));
        }
        private void modify() {
            int found = search();
            if (found < 0) {
                System.out.println("没有这个联系人");
                return;
            }
            list.remove(found);
            add();
        }
        private void delete() {
            int found = search();
            if (found >= 0) {
                list.remove(found);
                System.out.println("删除成功");
            } else
                System.out.println("没有这个联系人");
        }
        private int search() {
            System.out.print("请输入名称: ");
            int found = -1;
            String name = sc.nextLine();
            for (int i = 0; i < list.size(); i++) {
                if (name.equalsIgnoreCase(list.get(i).getName())) {
                    found = i;
                    break;
                }
            }
            return found;
        }
    }
```

修改 Main 类，创建 Manager 的对象，并调用 Manager 的 run( ) 方法运行。Main 类的代码如下（源代码为 11-11.java）:

```java
package org.example;
import com.ttt.contact.Manager;
public class Main {
    public static void main(String[] args) {
```

```
        Manager m = new Manager();
        m.run();
    }
}
```

运行这个程序，显示以下结果。

```
目前没有任何联系信息
Menu
1: 查询通讯录信息
2: 增加通讯录信息
3: 修改通讯录信息
4: 删除通讯录信息
0: 退出
请输入你的选择: 2
请输入名称: 张三
请输入电话号码: 1380013××××
请输入通讯录地址: 广州市
```

11.7 练习：自制工资管理程序

在工资管理程序中，一个人的基本信息包括姓名、工作单位、工资等。设计一个工资管理程序，可以实现工资信息的增加、删除、修改和查询。要求：将工资信息保存在文件中，并且，在启动程序时从文件中读取所有的工资信息，在退出程序时，将工资信息保存到文件中。

第 12 章　Java 反射和注解

Java 反射和注解机制是设计和编写高级程序的重要技术。Java 反射机制的一个典型应用就是：在 IDEA 编程环境中，当输入某个对象变量的名字时，IDEA 可以立即显示该对象包括的方法和属性供编码者选择，以减轻对编码者的记忆要求。Java 注解为编写高级程序框架提供了支撑。典型的 Spring 框架就大量使用了 Java 注解技术。本章对 Java 的反射和注解机制进行介绍。

12.1　Java 反 射

Java 反射机制，也称为 Java Reflection，是一种在没有 Java 源代码的情况下，从 Java 编译生成的 class 文件获取类的定义、创建类的对象、调用对象的方法的一种机制。简单来说就是：通过反射机制，可以从类的 .class 文件生成与这个类的源代码几乎一致的 Java 源代码。

12.1.1　反射概念的引入

为了对 Java 反射机制有一个直观了解，先看一个简单的例子。为此，在 IDEA 中新建一个名为 ch12-01 的 Java 工程。在工程中选择 External Libraries → <17> → java.base → com.sun → crypto.provider → AESCipher 命令，如图 12-1 所示。可以发现在右边的编辑窗口中显示了 AESCipher 类的源代码。

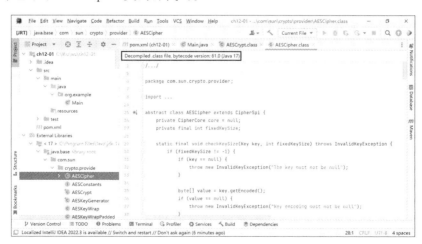

图 12-1　通过反射查看 class 的源代码

在图 12-1 中，注意矩形框中的代码，这段代码的意思是：从 .class 文件反编译出 Java 类源代码。当然，反射的应用远远不止这些。

IDEA 是如何做到从 Java 编译生成的 .class 文件反编译出 Java 类的源代码的呢？答案就是 Java 反射机制。Java 反射机制的核心是 Java 的 Class 类。注意，Class 是 JDK 定义的一个类，与之前用过的其他类一样，如 String 类。还要注意，Class 是 Java 的一个类，这里的 C 是大写的，不要与 class 关键字造成混淆。

12.1.2 反射的核心——Class 类

Java 虚拟机运行 Java 程序时，会为程序中用到的每个 Java 类创建一个全局唯一的 Class 对象，这个 Class 对象表征了每个类的所有信息。因此，每个程序类对应的 Class 对象对获取对应类的信息具有重要的作用。这些程序类可以是：JDK 中已经定义的类、第三方提供的类、编码者自己定义的类。有三种获取程序类所对应的 Class 对象的方式。

（1）类名 .class。也就是对于任何已经定义的类，可以直接通过这种方式得到这个类的 Class 对象。其中的类，既可以是任何已经定义的类，包括 JDK 定义的类，如 String 等，也可以是自己定义的类。

（2）实例名 .getClass( )。通过对象的实例名获取对应类的 Class 对象。

（3）Class.forName(String className)。通过类的全限定名获取该类的 Class 对象。

为了直观地理解这一过程，先看一个简单的例子。为此，在 ch12-01 工程下新建一个名为 com.ttt.reflection 的包，然后在这个包下新建名为 Student 的 Java 类。Student 类的代码如下（源代码为 12-01.java）：

```java
package com.ttt.reflection;
public class Student {
    private String name;
    private int age;
    private String school;
    public Student() {
    }
    public Student(String name, int age, String school) {
        this.name = name;
        this.age = age;
        this.school = school;
    }
    public void setName(String name) {
        this.name = name;
    }
    public void setAge(int age) {
        this.age = age;
    }
    public void setSchool(String school) {
        this.school = school;
```

```
    }
    @Override
    public String toString() {
        return "Student{" +
                "name='" + name + '\'' +
                ", age=" + age +
                ", school='" + school + '\'' +
                '}';
    }
}
```

现在修改 Main 类，在 main( ) 方法中以三种方式创建 Student 类的 Class 对象，也以三种方式创建 Integer 类的 Class 对象，然后显示信息。修改后的 Main 类代码如下所示（源代码为 12-02.java）。

```
package org.example;
import com.ttt.reflection.Student;
public class Main {
    public static void main(String[] args) {
        Class<?> clazz1 = Student.class;
        Student stu = new Student("张三", 10, "第一小学");
        Class<?> clazz2 = stu.getClass();
        Class<?> clazz3;
        try {
            clazz3 = Class.forName("com.ttt.reflection.Student");
        } catch (ClassNotFoundException e) {
            throw new RuntimeException(e);
        }
        System.out.println(clazz3);
        Class<?> clazz4 = Integer.class;
        Integer i = 1234;
        Class<?> clazz5 = i.getClass();
        Class<?> clazz6;
        try {
            clazz6 = Class.forName("java.lang.Integer");
        } catch (ClassNotFoundException e) {
            throw new RuntimeException(e);
        }
        System.out.println(clazz6);
    }
}
```

以上代码以三种方式分别对 Student 类和 String 类创建其对应的 Class 对象。运行这个程序，显示以下信息。

```
class com.ttt.reflection.Student
class java.lang.Integer
```

一旦得到了类的 Class 对象，就可以基于这个 Class 对象，通过反射机制获取对应类的所有信息，包括类的构造方法、属性和普通方法等。

12.1.3　通过反射获取类的构造方法、属性和普通方法

Class 类提供一系列用于获取类信息的方法。Class 类的常用方法如表 12-1 所示。

表 12-1　Class 类的常用方法

方　　法	功　　能
static Class<?> forName(String className)	返回 className 指定的类或接口所对应的 Class 对象。注意，className 必须是全限定名，也就是包括包名和类名
Constructor<T> getDeclaredConstructor(Class<?>... parameterTypes)	获取与给定参数类型匹配类的构造方法描述对象
Constructor<?>[] getDeclaredConstructors( )	获取类的所有构造方法描述对象
Field getDeclaredField(String name)	获取类所定义的指定 name 名称的属性描述对象
Field[] getDeclaredFields( )	获取类的所有属性描述对象
Method getDeclaredMethod(String name, Class<?>... parameterTypes)	获取类所定义的与给定参数匹配的方法描述对象
Method[] getDeclaredMethods( )	获取类所定义的所有方法描述对象

从表 12-1 可以看出，通过类的 Class 对象可以获取类所定义的构造方法描述对象 Constructor、属性描述对象 Field、方法描述对象 Method。Constructor 类、Field 类、Method 类的常用方法分别如表 12-2~ 表 12-4 所示。

表 12-2　Constructor 类的常用方法

方　　法	功　　能
String getName( )	返回构造方法的名字
int getModifiers( )	返回构造方法的访问限定符
Class<?>[] getParameterTypes( )	返回一个描述构造方法参数类型的 Class 类型数组
T newInstance(Object... initargs)	用给定的参数，创建 Class 类所描述类的一个对象

表 12-3　Field 类的常用方法

方　　法	功　　能
String getName( )	返回属性的名字
int getModifiers( )	返回属性的访问限定符
Object get(Object obj)	返回 obj 对象中该属性的值对象
int getInt(Object obj)	如果该属性是 int 类型，返回该属性的 int 类型的值。类似地，还有返回其他基本数据类型的 getXXX( ) 方法
void set(Object obj, Object value)	设置 obj 对象中该属性的值为 value

续表

方法	功能
void setInt(Object obj, int i)	如果该属性是 int 类型，则设置 obj 对象的值为 int 类型参数 i 的值
boolean isAnnotationPresent(Class<? extends Annotation> annotationClass)	判断属性是否被 annotationClass 注解所标注，如是则返回 true，否则返回 false
<T extends Annotation> T getAnnotation (Class<T> annotationClass)	如果属性被 annotationClass 注解所标注，则返回注解对象；否则返回 null

表 12-4 Method 类的常用方法

方法	功能
String getName( )	返回方法的名字
int getModifiers( )	返回方法的访问限定符
Class<?> getReturnType( )	返回方法的返回值类型
Class<?>[] getParameterTypes( )	返回方法的参数类型数组
Object invoke(Object obj, Object... args)	用给定的参数调用该方法
boolean isAnnotationPresent(Class<? extends Annotation> annotationClass)	判断方法是否被 annotationClass 注解所标注，如是则返回 true，否则返回 false
<T extends Annotation> T getAnnotation (Class<T> annotationClass)	如果方法被 annotationClass 注解所标注，则返回注解对象；否则返回 null

下面举一个例子说明如何通过反射机制创建类的对象和调用对象的方法。这个例子首先获取 Student 类的 Class 对象，然后通过 Class 对象创建 Student 类的对象、调用普通方法设置属性的值，最后显示所创建 Student 对象的信息。为此，修改 Main 类，在 main( ) 方法中完成这些功能。修改后的 Main 类代码如下（源代码为 12-03.java）：

```java
package org.example;
import com.ttt.reflection.Student;
import java.lang.reflect.Constructor;
import java.lang.reflect.Field;
import java.lang.reflect.InvocationTargetException;
import java.lang.reflect.Method;
public class Main {
    public static void main(String[] args) {
        Class<Student> clazz = Student.class;
        Constructor<Student>[] cons = (Constructor<Student>[])
                                clazz.getDeclaredConstructors();
        System.out.println("Student 类有 " + cons.length + " 个构造方法 ");
        try {
            Constructor<Student> con = clazz.getDeclaredConstructor
            (String.class,int.class, String.class);
            Student stu = con.newInstance(" 张三 ", 10, " 第一小学 ");
            System.out.println(stu);
            Field fn = clazz.getDeclaredField("name");
```

```java
                System.out.println("name 属性的类型: " + fn.getType());
                Method m = clazz.getDeclaredMethod("setName", String.class);
                m.invoke(stu, "李四");
                System.out.println(stu);
        } catch (NoSuchMethodException e) {
            throw new RuntimeException(e);
        } catch (InvocationTargetException e) {
            throw new RuntimeException(e);
        } catch (InstantiationException e) {
            throw new RuntimeException(e);
        } catch (IllegalAccessException e) {
            throw new RuntimeException(e);
        } catch (NoSuchFieldException e) {
            throw new RuntimeException(e);
        }
    }
}
```

在这个程序中，首先使用语句：

```java
Class<Student> clazz = Student.class;
```

创建了 Student 类的 Class 对象，然后使用语句：

```java
Constructor<Student>[] cons = (Constructor<Student>[])
                                    clazz.getDeclaredConstructors();
System.out.println("Student 类有 " + cons.length + " 个构造方法 ");
```

获取 Student 类的所有构造方法描述对象并显示构造方法的数目。再使用语句：

```java
Constructor<Student> con = clazz.getDeclaredConstructor(String.
class,int.class, String.class);
Student stu = con.newInstance(" 张三 ", 10, " 第一小学 ");
```

获取指定的构造方法描述对象，并基于构造方法创建 Student 类的一个对象。又使用语句：

```java
Field fn = clazz.getDeclaredField("name");
System.out.println("name 属性的类型: " + fn.getType());
```

获取 name 属性的描述对象，并显示 name 属性的类型名。接着使用语句：

```java
Method m = clazz.getDeclaredMethod("setName", String.class);
m.invoke(stu, "李四");
```

获取 setName( ) 方法的描述对象,并调用这个方法设置 stu 对象的 name 属性为 "李四"。最后显示 stu 对象的信息。运行这个程序,显示以下信息。

```
Student 类有 2 个构造方法
Student{name='张三', age=10, school='第一小学'}
name 属性的类型: class java.lang.String
Student{name='李四', age=10, school='第一小学'}
```

从运行结果发现,可以使用反射获取 Student 类的构造方法、属性对象描述,并使用反射得到的方法设置 stu 对象的 name 属性值。

12.2 Java 注 解

Java 注解,也称为 Java 标注,英文名称为 Annotation,是 Java 提供的一种高级编程机制,它是实现 Java 框架编程的重要支撑手段。例如,著名的 Spring 框架就大量使用了 Java 注解。本节对 Java 注解进行介绍。

Java 注解是附加在代码中的一些元信息,用于一些工具在编译、运行时进行解析和使用,起到说明、配置的功能。例如,使用 @Override 注解注解类的某个方法,Java 编译器在编译 Java 源代码时检查所标注的方法是否重写了父类的同名方法,如果父类中不存在同名的方法,则编译器会报错;@FunctionInterface 注解用于注解一个接口,要求所注解的接口有且只能有一个抽象方法,如果接口不满足这个要求,Java 编译器在编译这个接口源代码时就会报错。

Java 注解的本质

12.2.1 Java 标准注解

Java 提供了一些注解,用以告知编译器所编译的 Java 源代码的一些特征,这些注解称为 Java 标准注解。Java 的常用标准注解如表 12-5 所示。

表 12-5 Java 的常用标准注解

注 解	作 用
@Override	编译器检查用该注解标注的方法是否为重写方法,如果其父类或者接口中没有定义该方法,编译器会报错
@FunctionInterface	用该注解所标注的接口是否有且只有一个抽象方法,如果不是,编译器会报错
@Deprecated	用该注解标注的方法已经过时,不应该使用这个方法,如果使用了用该注解标注的方法,编译器会告警
@SuppressWarnings	指示编译器忽略注解中出现的警告

12.2.2 自定义注解

Java 的编码者可以根据需要自定义注解,并通过自定义注解完成某些特定的功能。著名的 Spring 框架就自定义了一系列注解以完成特定功能。为了自定义注解，Java 提供了一些称为元注解的最为基本的注解，为编码者自定义注解提供支撑。Java 提供的元注解如表 12-6 所示。

表 12-6　Java 提供的元注解

元注解	作用
@Documented	用以指示文档生成工具，将该代码的文档纳入其生成的文档中，无参数
@Retention	定义该注解的生命周期。有一个参数，参数可取值及其含义如下。 （1）RetentionPolicy.SOURCE：这类注解指示编译器的行为，在编译完成之后丢弃，注解不会被写入字节码 class 文件中。 （2）RetentionPolicy.CLASS：这类注解指示类加载器的行为，在类加载完成之后丢弃，在字节码文件的处理中有用。注解默认使用这种方式。 （3）RetentionPolicy.RUNTIME：始终不会丢弃，运行期也保留该注解，因此可以使用反射机制读取该注解的信息
@Target	表示该注解可用于标注哪些元素。有一个参数，参数可取值及其含义如下。 （1）ElementType.CONSTRUCTOR：用于描述构造器。 （2）ElementType.FIELD：用于描述成员变量、对象、属性（包括 enum 实例）。 （3）ElementType.LOCAL_VARIABLE：用于描述局部变量。 （4）ElementType.METHOD：用于描述方法。 （5）ElementType.PACKAGE：用于描述包。 （6）ElementType.PARAMETER：用于描述参数。 （7）ElementType.TYPE：用于描述类、接口（包括注解类型）或 enum 声明
@Inherited	指明该注解可以被继承。没有参数

创建自定义注解和创建一个接口相似，但是注解的 interface 关键字需要以 @ 符号开头，可以为注解声明方法。自定义注解的一般格式如下：

```
元注解
...
public @interface 注解名称 {
    // 属性列表
}
```

下面举例说明如何自定义注解及如何使用自定义注解。这个例子首先定义名为 @CompanyName 的注解，然后使用这个注解标注 Fruit 类的一个属性。继而编写一个从 Fruit 的 Class 对象获取注解值的工具类 MyAnnotationUtil，然后，在 Main 类中，创建 Fruit 类的对象，获取使用 @CompanyName 注解标注的属性，并设置这个属性的属性值为标注的参数值。最后，显示 Fruit 对象的信息。@CompanyName 注解的代码如下（源代码为 12-04.java）：

```java
package com.ttt.annotation;
import java.lang.annotation.Retention;
import java.lang.annotation.Target;
import static java.lang.annotation.ElementType.FIELD;
import static java.lang.annotation.RetentionPolicy.RUNTIME;
```

```
@Target(FIELD)
@Retention(RUNTIME)
public @interface CompanyName {
    String name() default " 大发水果公司 ";
}
```

注意代码中的这条语句:

```
import static java.lang.annotation.ElementType.FIELD;
import static java.lang.annotation.RetentionPolicy.RUNTIME;
```

它们"静态导入"所需类的指定成员。使用静态导入可以从指定类中导入类的静态成员，包括静态属性和静态方法。使用静态导入语句导入了类的静态成员后，后续只需要使用成员名即可使用已经导入的静态成员，而不再需要加上类的名称。在导入所需的类后，使用语句:

```
@Target(FIELD)
@Retention(RUNTIME)
public @interface CompanyName {
    String name() default " 大发水果公司 ";
}
```

定义了名为 CompanyName 的标注。注意，由于前面已经静态导入了 FIELD 和 RUNTIME 属性，因此，在这里可以直接使用这两个属性名。

Fruit 类的代码如下（源代码为 12-05.java）:

```
package com.ttt.annotation;
public class Fruit {
    @CompanyName(name=" 某水果销售有限公司 ")
    public String company;
    private String name;
    private float price;
    public Fruit() {
    }
    public Fruit(String name, float price) {
        this.name = name;
        this.price = price;
    }
    public String getName() {
        return name;
    }
    public void setName(String name) {
        this.name = name;
```

```
    }
    public float getPrice() {
        return price;
    }
    public void setPrice(float price) {
        this.price = price;
    }
    @Override
    public String toString() {
        return "Fruit{" +
                "name='" + name + '\'' +
                ", price=" + price +
                '}' + ", 由" + "'" + company + "'" + "荣誉出品";
    }
}
```

Fruit 类指示一个简单的 POJO 类，比较简单。在 Fruit 类中，使用语句：

```
@CompanyName(name=" 某水果销售有限公司 ")
public String company;
```

对属性 company 进行了标注。

MyAnnotationUtil 工具类的代码如下（源代码为 12-06.java）:

```
package com.ttt.annotation;
import java.lang.reflect.Field;
public class MyAnnotationUtil {
    public static void getCompanyName(Fruit fruit) {
        Field[] fields = fruit.getClass().getDeclaredFields();
        for(Field field :fields) {
            if (field.isAnnotationPresent(CompanyName.class)) {
                CompanyName cn = field.getAnnotation(CompanyName.class);
                try {
                    field.set(fruit, cn.name());
                } catch (IllegalAccessException e) {
                    throw new RuntimeException(e);
                }
            }
        }
    }
}
```

在 MyAnnotationUtil 工具类的 getCompanyName( ) 静态方法中，首先使用反射机制得到 Fruit 类所定义的所有属性的 Field 描述对象。

```
Field[] fields = fruit.getClass().getDeclaredFields();
```

然后使用以下语句对所有的属性进行遍历。

```
if (field.isAnnotationPresent(CompanyName.class)) {
    CompanyName cn = field.getAnnotation(CompanyName.class);
    try {
        field.set(fruit, cn.name());
    } catch (IllegalAccessException e) {
        throw new RuntimeException(e);
    }
}
```

检查属性是否被 @CompanyName 所标注，若是，则获取标注对象，得到标注的值，并将得到的值设置到对应的属性中。

最后，修改 Main 类为以下代码（源代码为 12-07.java）。

```
package org.example;
import com.ttt.annotation.Fruit;
import com.ttt.annotation.MyAnnotationUtil;
public class Main {
    public static void main(String[] args) {
        Fruit f = new Fruit("苹果", 15.5F);
        MyAnnotationUtil.getCompanyName(f);
        System.out.println(f);
    }
}
```

在 Main 类中，先创建 Fruit 类的对象，再使用工具类 MyAnnotationUtil 的静态方法 getCompanyName( ) 检查 Fruit 对象并设置被 @CompanyName 所标注属性的值，并显示 Fruit 对象的信息。运行这个程序，显示以下信息。

```
Fruit{name='苹果', price=15.5}, 由'某水果销售有限公司'荣誉出品
```

从运行结果可以看出，Fruit 对象的 company 属性值被设置为 @CompanyName 注解的参数值。

12.3 Java 反射与注解应用实践

作为一般的程序设计人员或编码人员，在编程实践中很少需要设计处理 Java 反射与注解的代码：对反射机制只需要了解；对注解机制只需要能够使用已经定义好的注解

即可，如可以灵活使用 @Override 注解、@FunctionInterface 等即可。之所以在本章介绍 Java 的反射和注解，有以下两点原因。

（1）在未来使用 Java 进行 Web 应用开发，特别是使用 Spring 框架进行应用开发实践时，会大量使用 Spring 已经定义的注解。如果不了解 Java 反射和注解机制的技术原理，不能知其所以然。

（2）反射和注解是就业面试时经常被问及的技术问题。

总之，学习 Java 的反射和注解是有意义的。当然，如果未来需要编写高质量的 Java 框架代码，需要更深入地学习 Java 的反射和注解机制。

12.4　案例：自动注入对象

自动注入对象，在 Spring 中被称为 Bean 注入，是非常基础、非常重要的技术。本节通过一个简单的案例强化对 Java 反射和注解的学习。

12.4.1　案例任务

使用 Java 反射和注解可以编写高质量的程序。编写一个简单的 Java POJO 类 Student，然后编写一个 Manager 类，其中包含一个 Student 属性，要求使用反射和注解机制向 Manager 中注入 Student 对象。然后编写一个测试类验证代码的正确性。

12.4.2　任务分析

Student 是一个 POJO 类，代码相对比较简单。对于 Manager 类，按要求，需要包含一个 Student 属性，而 Student 属性的对象需要采用反射和注解方式自动注入。为了实现这个要求，需要编写一个注解，对需要注入 Student 的属性进行标注。当然，在 Manager 类中还需要定义其他一些用于 Student 属性的方法。总结起来，需要编写以下几类。

（1）Student 类：POJO 类。

（2）Manager 类：其中包含需要采用反射和注解机制自动注入 Student 对象的属性，当然，在 Manager 类中还需要包含其他一些方法。

（3）AutoWired 注解：用于标注需要注入 Student 对象的其他类的属性。

（4）MyAnnotationUtil 类：一个工具类，用于分析目标对象，并自动为目标对象注入 Student 对象。

（5）Main 类：用于测试代码的正确性。

12.4.3　任务实施

在 ch12-01 工程下新建一个名为 com.ttt.auto 的包，在其中新建 Studnet 类、Manager 类、AutoWired 注解和 MyAnnotationUtil 类。Student 类的代码如下（源代码为 12-08.java）：

```java
package com.ttt.auto;
public class Student {
    private String name;
    private int age;
    private String school;
    public Student(String name, int age, String school) {
        this.name = name;
        this.age = age;
        this.school = school;
    }
    public String getName() {
        return name;
    }
    public void setName(String name) {
        this.name = name;
    }
    public int getAge() {
        return age;
    }
    public void setAge(int age) {
        this.age = age;
    }
    public String getSchool() {
        return school;
    }
    public void setSchool(String school) {
        this.school = school;
    }
    @Override
    public String toString() {
        return "Student{" +
                "name='" + name + '\'' +
                ", age=" + age +
                ", school='" + school + '\'' +
                '}';
    }
}
```

AutoWired 注解的代码如下（源代码为 12-09.java）：

```java
package com.ttt.auto;
import java.lang.annotation.Retention;
import java.lang.annotation.Target;
import static java.lang.annotation.ElementType.FIELD;
```

```
import static java.lang.annotation.RetentionPolicy.RUNTIME;
@Target(FIELD)
@Retention(RUNTIME)
public @interface AutoWired {
}
```

Manager 类的代码如下（源代码为 12-10.java）：

```
package com.ttt.auto;
public class Manager {
    @AutoWired
    private Student student;
    private final String who;
    public Manager(String who) {
        this.who = who;
    }
    public void setStudent(Student student) {
        this.student = student;
    }
    public void ageIncOne() {
        student.setAge(student.getAge() + 1);
    }
    public void displayStudent() {
        System.out.println(student);
    }
}
```

MyAnnotationUtil 类的代码如下（源代码为 12-11.java）：

```
package com.ttt.auto;
import java.lang.reflect.Field;
import java.lang.reflect.InvocationTargetException;
import java.lang.reflect.Method;
public class MyAnnotationUtil {
    private static Student student = new Student("张三", 10, "第二小学");
    public static void autowire_student(Manager manager) {
        Class<?> clazz = manager.getClass();
        Field[] fields = clazz.getDeclaredFields();
        for(Field field :fields) {
            if (field.isAnnotationPresent(AutoWired.class)) {
                try {
                    Method method = clazz.getDeclaredMethod("setStudent",
                    Student.class);
                    method.invoke(manager, student);
```

```
            } catch (NoSuchMethodException e) {
                throw new RuntimeException(e);
            } catch (InvocationTargetException e) {
                throw new RuntimeException(e);
            } catch (IllegalAccessException e) {
                throw new RuntimeException(e);
            }
        }
    }
}
```

Main 类的代码如下（源代码为 12-12.java）：

```
package org.example;
import com.ttt.auto.MyAnnotationUtil;
import com.ttt.auto.Manager;
public class Main {
    public static void main(String[] args) {
        Manager man = new Manager("管理员");
        MyAnnotationUtil.autowire_student(man);
        man.ageIncOne();
        man.displayStudent();
    }
}
```

运行这个程序，显示以下结果。

```
Student{name='张三', age=11, school='第二小学'}
```

结果是正确的。

12.5 练习：自动注入 Teacher 对象

使用 Java 反射和注解可以编写高质量的程序。首先编写一个简单的 Java POJO 类 Teacher，然后编写一个 TeacherManager 类，其中包含一个 Teacher 属性，要求使用反射和注解机制向 TeacherManager 中注入 Teacher 对象。最后编写一个测试类验证代码的正确性。

第 13 章　多　线　程

现代 CPU 一般都是多核 CPU。所谓多核，是指在一颗 CPU 芯片中有多个可执行运算任务的核心单元，并且一个核心单元可执行一个或多个运算任务。为了与 CPU 的多核技术同步，现代编程语言一般也都支持多线程编程：将一个程序划分为包含多个完成特定任务功能的执行单元，这些执行单元相互协同以完成程序的整体功能。程序中每个完成特定任务功能的执行单元称为线程。匹配于 CPU 的多核技术，可将程序中不同线程安排到不同的 CPU 核心去执行，从而在有效提升 CPU 利用率的同时也提高了程序的执行效率。Java 语言对多线程编程提供了非常好的支持。本章对 Java 的多线程编程进行介绍。

13.1　Java 多线程入门

在本章之前编写的所有例子程序都是只包含一个线程的程序，程序所包含的那个唯一线程称为 main 线程：因为这个线程是在 Java 虚拟机执行程序时，通过调用程序静态的 main( ) 方法而创建的。main 线程有时也称为主线程。在深入讲解 Java 多线程技术之前，先看一个简单的 Java 多线程例子。

这个例子首先创建一个线程用于显示当前日期时间的子线程：每隔 1500 毫秒显示一次日期时间信息。同时，在 main 主线程中每隔 500 毫秒显示"这是主线程在运行 …"的信息。为此，在 IDEA 中新建一个名为 ch13-01 的 Java 工程，完成后的 IDEA 界面如图 13-1 所示。

图 13-1　新建 ch13-01 工程后的 IDEA 界面

然后，修改 Main 类为以下代码（源代码为 13-01.java）。

```
package org.example;
import java.text.SimpleDateFormat;
```

```java
import java.util.Date;
public class Main {
    public static void main(String[] args) {
        System.out.println("Creating Timer Thread...");
        PrintTimerThread ptt = new PrintTimerThread();
        System.out.println("Start thread...");
        ptt.start();
        for(int i=0; i<10; i++) {
            System.out.println("这是主线程在运行...");
            try {
                Thread.sleep(500);
            } catch (InterruptedException e) {
                throw new RuntimeException(e);
            }
        }
        System.out.println("Over!");
    }
    private static class PrintTimerThread extends Thread {
        @Override
        public void run() {
            long current = System.currentTimeMillis();
            SimpleDateFormat sdf = new SimpleDateFormat("yyyy-MM-dd HH:mm:ss");
            while (System.currentTimeMillis() - current < 10*1000) {
                Date d = new Date();
                System.out.println("当前日期时间是:" + sdf.format(d));
                try {
                    Thread.sleep(1500);
                } catch (InterruptedException e) {
                    throw new RuntimeException(e);
                }
            }
        }
    }
}
```

在 Main 类中，首先定义了一个继承自 Thread 父类的 PrintTimerThread 子类，并重写了 Thread 的 run( ) 方法。在 run( ) 方法中，通过一个循环在屏幕上显示当前的日期时间信息，之后调用语句：

```
Thread.sleep(1500);
```

使线程睡眠 1500 毫秒，之后继续循环直到结束。

定义了 PrintTimerThread 类之后，在 Main 类的 main( ) 方法中使用语句：

```
PrintTimerThread ptt = new PrintTimerThread();
```

创建了 PrintTimerThread 类的对象，也就是创建了一个线程，这个线程也称为子线程。注意，这里只是创建了线程对象，子线程并没有开始运行。为了启动子线程运行，使用语句：

```
ptt.start();
```

此时，线程才真正开始运行，屏幕上会显示当前日期时间信息。

如前所述，Java 虚拟机通过调用 main( ) 方法启动 Java 程序运行后，main 主线程也开始运行了。为了使在子线程运行期间主线程仍然运行，在 main( ) 方法中编写一段循环代码。

```
for(int i=0; i<10; i++) {
    System.out.println("这是主线程在运行...");
    try {
        Thread.sleep(500);
    } catch (InterruptedException e) {
        throw new RuntimeException(e);
    }
}
```

这段代码在一个循环中每隔 500 毫秒不断显示"这是主线程在运行..."这段文字信息。现在运行这个程序，显示以下信息。

```
Creating Timer Thread...
Start thread...
这是主线程在运行...
当前日期时间是：2023-01-25 07:40:09
这是主线程在运行...
这是主线程在运行...
这是主线程在运行...
...// 省略了一些显示信息
当前日期时间是：2023-01-25 07:40:13
Over!
当前日期时间是：2023-01-25 07:40:15
当前日期时间是：2023-01-25 07:40:16
当前日期时间是：2023-01-25 07:40:18
```

从运行结果可以看出，main 主线程和打印日期时间的线程都在运行。主线程每隔 500 毫秒显示一次"这是主线程在运行..."信息，而子线程每隔 1500 毫秒显示一次日期时间信息。同时，留意输出结果中的"Over!"文字信息：这是主线程结束之前显示的最后信息，一旦显示了这个信息，表示主线程已经运行结束。而此时子线程并没有结束，所以，还会继续显示三条日期时间信息。

从这个例子可以看出,Thread 类在 Java 多线程中起着重要作用。下面介绍 Thread 类及如何通过 Thread 类创建新线程。

13.2 Thread 类及创建子线程

Thread 类是 Java 实现多线程编程的主要支持类。在 Java 中,线程都是 Thread 类或其子类的对象。Thread 类的常用方法如表 13-1 所示。

表 13-1 Thread 类的常用方法

方　　法	功　　能
Thread( )	创建一个线程对象
Thread(String name)	创建一个线程对象,设置该线程的名字为 name
Thread(Runnable target)	使用 target 参数创建一个线程对象。Runnable 是函数式接口,其中有一个 run( ) 方法
Thread(Runnable target, String name)	使用 target 参数创建一个线程对象,并设置该线程的名字为 name
static Thread currentThread( )	返回正在运行的线程对象
static int enumerate(Thread[] tarray)	获取正在运行的所有线程对象到 tarray 数组中
long getId( )	返回线程的 id
final String getName( )	返回线程的名字
final int getPriority( )	返回线程的优先级
Thread.State getState( )	返回线程的状态。其中 Thread.State 是枚举类型,包括以下枚举值:NEW、RUNNABLE、BLOCKED、WAITING、TIMED_WAITING、TERMINATED
void interrupt( )	中断线程执行
final void join( )	等待线程运行结束
final void join(long millis, int nanos)	在指定的时长内等待线程结束
void run( )	当线程被启动时,将自动调用这个方法启动线程任务执行。注意,Thread 定义的 run( ) 方法没有执行任何功能,Thread 的子类需要覆盖这个方法来完成线程任务
final void setName(String name)	设置线程的名称
final void setPriority(int newPriority)	设置线程的运行优先级
static void sleep(long millis)	使调用这个函数的线程休眠指定的毫秒
void start( )	启动线程开始执行
static void yield( )	线程主动放弃 CPU,然后等待下一次被调度执行

从表 13-1 可以看出,Thread 类有 4 个常用的构造方法,其中 Thread( )、Thread (String name) 用于在派生子类中使用,并在派生的子类中重写 run( ) 方法,这个重写的 run( ) 方法用于完成线程的任务;而 Thread(Runnable target) 和 Thread(Runnable target, String name) 则可用于直接构造 Thread 类的对象,传入的 target 参数用于完成线程的任务。

13.2.1 通过继承 Thread 类创建线程

第 13.1 节的例子就是通过定义 Thread 类的子类，并在子类中重写 run( ) 方法来完成线程任务的。关于这部分内容，此处不再赘述。

13.2.2 通过实现 Runnable 接口创建线程

先实现 Runnable 接口，再通过 Thread 的构造函数 Thread(Runnable target) 或者 Thread(Runnable target) 来创建线程是非常常用和便利的创建线程的方式。Runnable 接口是一个函数式接口，只有一个抽象方法 run( )，它的原型是 void run( )。下面举例说明如何使用 Thread(Runnable target, String name) 来创建线程。

修改 Main 类，直接使用 Thread 类创建线程。这个线程完成的功能与第 13.1 节的例子完全一致。修改后的 Main 类代码如下（源代码为 13-02.java）：

```java
package org.example;
import java.text.SimpleDateFormat;
import java.util.Date;
public class Main {
    public static void main(String[] args) {
        System.out.println("Creating Timer Thread...");
        Thread ptt = new Thread(new Runnable() {
            @Override
            public void run() {
                long current = System.currentTimeMillis();
                SimpleDateFormat sdf = new SimpleDateFormat("yyyy-MM-dd HH:mm:ss");
                while (System.currentTimeMillis() - current < 10*1000) {
                    Date d = new Date();
                    System.out.println("当前日期时间是: " + sdf.format(d));
                    try {
                        Thread.sleep(1500);
                    } catch (InterruptedException e) {
                        throw new RuntimeException(e);
                    }
                }
            }
        }, "一个打印线程");
        System.out.println("Start thread ...");
        ptt.start();
        for(int i=0; i<10; i++) {
            System.out.println("这是主线程在运行...");
```

```
            try {
                Thread.sleep(500);
            } catch (InterruptedException e) {
                throw new RuntimeException(e);
            }
        }
        System.out.println("Over!");
    }
}
```

运行这个程序，显示与第 13.1 节例子完全一致的结果。

每个线程对象都有名称、id、所处的状态和线程执行优先级。在创建一个线程对象后，可以调用相应方法获取这些信息。例如，在上面所示的例子中，在创建了 ptt 线程后，可以使用以下代码获取线程的相关信息。

```
System.out.println("线程名称: " + ptt.getName() + "\n" +
                   "线程id: " + ptt.getId() + "\n" +
                   "线程状态: " + ptt.getState() + "\n" +
                   "线程优先级: " + ptt.getPriority());
```

在 Thread 类中，有一个非常重要的方法：join( ) 方法。这个方法用于等待指定的线程结束运行。为了理解 join( ) 方法的作用，修改上面的例子，增加显示线程信息的语句，同时删除 main 主线程中循环打印信息的代码片段，代之以对 ptt 线程 join( ) 方法的调用。修改后的代码如下（源代码为 13-03.java）:

```
package org.example;
import java.text.SimpleDateFormat;
import java.util.Date;
public class Main {
    public static void main(String[] args) {
        System.out.println("Creating Timer Thread...");
        Thread ptt = new Thread(new Runnable() {
            @Override
            public void run() {
                long current = System.currentTimeMillis();
                SimpleDateFormat sdf = new SimpleDateFormat("yyyy-MM-dd HH:mm:ss");
                while (System.currentTimeMillis() - current < 10*1000) {
                    Date d = new Date();
                    System.out.println("当前日期时间是: " + sdf.format(d));
                    try {
                        Thread.sleep(1500);
                    } catch (InterruptedException e) {
```

```
                    throw new RuntimeException(e);
                }
            }
        }
    }, "一个打印线程");
    System.out.println("线程名称:" + ptt.getName() + "\n" +
                       "线程id:" + ptt.getId() + "\n" +
                       "线程状态:" + ptt.getState() + "\n" +
                       "线程优先级:" + ptt.getPriority());
    System.out.println("Start thread...");
    ptt.start();
    try {
        ptt.join();
    } catch (InterruptedException e) {
        throw new RuntimeException(e);
    }
    System.out.println("Over!");
    }
}
```

运行这个程序，显示以下信息。

```
Creating Timer Thread ...
线程名称:一个打印线程
线程id:16
线程状态:NEW
线程优先级:5
Start thread ...
当前日期时间是:2023-01-25 10:08:10
...// 省略了一些显示信息
当前日期时间是:2023-01-25 10:08:18
当前日期时间是:2023-01-25 10:08:19
Over!
```

程序正确获取线程的相关信息，包括名称、id、线程状态和线程优先级。同时，当子线程运行完毕并显示了所有的日期时间信息后，main 主线程才在最后时刻显示 "Over!" 信息，表示主线程运行结束。因此，join( ) 方法的作用就是：等待指定线程结束运行。例如，在上面这个例子中，主线程中调用以下语句：

```
ptt.join();
```

等待子线程 ptt 结束运行。待子线程 ptt 结束运行后，主线程继续运行，因此，主线程打印出 "Over!" 文字信息。

13.2.3 使用 FutureTask 创建线程

在编程实践中,有时需要获取线程的运行结果,或者,需要取消某个正在运行的线程继续运行。然而,第 13.2.1 小节介绍的通过继承 Thread 类创建线程和第 13.1.2 小节介绍的通过实现 Runnable 接口创建线程两种方式,均不能满足这些要求。为此,Java 提供了 FutureTask 类来创建可以有返回值的线程。

FutureTask 类同时实现了 Runnable 接口和 Future 接口:因为实现了 Runnable 接口,所以可以作为 Thread 构造方法的参数;又因为实现了 Future 接口,所以可以将线程的执行结果保存在 Future 中以备后续需要或者取消线程继续执行。FutureTask 类的常用方法如表 13-2 所示。

表 13-2 FutureTask 类的常用方法

方 法	功 能
FutureTask(Callable\<V\> callable)	基于 callable 参数创建 FutureTask 对象。线程启动后将执行 callable 中的 call() 方法。Callable 是函数式接口,有一个 call() 抽象方法,其原型是 V call()。V 是 FutureTask 的泛型参数
FutureTask(Runnable runnable, V result)	基于 runnable 和 result 创建 FutureTask 对象,其中 runnable 是线程启动时被执行的方法,而 result 则是线程结束运行后的返回值。V 是 FutureTask 的泛型参数
boolean cancel(boolean mayInterruptIfRunning)	取消线程继续执行
boolean isCancelled( )	检查线程是否已被取消执行
V get( )	等待线程执行完成,当线程执行结束时,获取线程的返回值。V 是 FutureTask 的泛型参数
V get(long timeout, TimeUnit unit)	在指定的时间内等待线程执行完成,当线程执行结束时,获取线程的返回值;否则返回 null。V 是 FutureTask 的泛型参数
boolean isDone( )	检查线程是否已经执行完成

下面举例说明如何使用 FutureTask 创建线程。这个例子创建两个子线程:一个子线程每个 1 秒打印一段日期时间信息,采用实现 Runnable 接口方式创建这个线程;另一个子线程随机寻找任何一个在 1000000 到 9000000 之间的水仙花数,采用 FutureTask 创建这个线程。主线程则等待两个线程运行结束,并显示得到的水仙花数。所谓水仙花数,是指一个 n 位数,其各个数位上数的 n 次方之和恰好等于原数。为此,在 ch13-01 工程下新建名为 com.ttt.future 的包,在这个包下新建 MyFuture 类。MyFuture 类的代码如下(源代码为 13-04.java):

```
package com.ttt.future;
import java.text.SimpleDateFormat;
import java.util.Date;
import java.util.Random;
import java.util.concurrent.Callable;
```

```java
import java.util.concurrent.FutureTask;
public class MyFuture {
    private FutureTask<Integer> fti;
    public MyFuture() {
    }
    public Thread createFirstThread() {
        Thread t1 = new Thread(new Runnable() {
            final SimpleDateFormat sdf = new SimpleDateFormat("yyyy-MM-dd HH:mm:ss");
            final Date d = new Date();
            @Override
            public void run() {
                int i = 0;
                while(i<5) {
                    d.setTime(System.currentTimeMillis());
                    System.out.println(sdf.format(d));
                    i++;
                    try {
                        Thread.sleep(1000);
                    } catch (InterruptedException e) {
                        throw new RuntimeException(e);
                    }
                }
            }
        }, "打印日期时间线程");
        return t1;
    }
    public Thread createSecondThread() {
        fti = new FutureTask<>(new Callable<Integer>() {
            final Random random = new Random(System.currentTimeMillis());
            @Override
            public Integer call() throws Exception {
                while(true) {
                    int r = random.nextInt(1000000, 9000000);
                    if (isNarcissistic(r))
                        return r;
                }
            }
        });
        Thread t2 = new Thread(fti);
        return t2;
    }
    public FutureTask<Integer> getFti() {
        return fti;
    }
}
```

```java
    private boolean isNarcissistic(int d) {
        String sd = "" + d;
        String[] digits = sd.split("");
        int dd = 0;
        for (String digit : digits) {
            dd += Math.pow(Integer.parseInt(digit), digits.length);
        }
        return dd == d;
    }
}
```

代码中的 createFirstThread( ) 方法用于创建每隔 1 秒显示日期时间的线程，采用实现 Runnable 接口方式创建该线程；而 createSecondThread( ) 则是创建寻找水仙花数的线程。采用以下代码。

```java
fti = new FutureTask<>(new Callable<Integer>() {
    final Random random = new Random(System.currentTimeMillis());
    @Override
    public Integer call() throws Exception {
        while(true) {
            int r = random.nextInt(1000000, 9000000);
            if (isNarcissistic(r))
                return r;
        }
    }
});
Thread t2 = new Thread(fti);
```

基于 Callable 接口创建了 FutureTask 对象，并基于得到的 FutureTask 对象创建了 Thread 线程。再修改 Main 类为以下代码（源代码为 13-05.java）。

```java
package org.example;
import com.ttt.future.MyFuture;
import java.util.concurrent.ExecutionException;
public class Main {
    public static void main(String[] args) {
        MyFuture mf = new MyFuture();
        Thread t1 = mf.createFirstThread();
        Thread t2 = mf.createSecondThread();
        t1.start();
        t2.start();
        try {
            int narcissistic = mf.getFti().get();
            System.out.println("水仙花数: " + narcissistic);
```

```
        } catch (InterruptedException | ExecutionException e) {
            throw new RuntimeException(e);
        }
        try {
            t1.join();
        } catch (InterruptedException e) {
            throw new RuntimeException(e);
        }
    }
}
```

在 Main 类的代码中，使用以下代码：

```
int narcissistic = mf.getFti().get();
System.out.println("水仙花数: " + narcissistic);
```

获取线程执行的结果并显示。运行这个程序，显示以下结果。

```
2023-01-26 10:49:42
水仙花数：1741725
2023-01-26 10:49:43
2023-01-26 10:49:44
2023-01-26 10:49:45
2023-01-26 10:49:46
```

13.3 线程状态、线程调度和线程优先级

线程是有状态的。例如，一个新建的线程对象和一个正在运行的线程对象显然处于不同的状态。线程有 6 种可能的状态，它们的名称及含义如下。

（1）NEW：一个新创建的并且还没有启动的线程处于这个状态。
（2）RUNNABLE：一个正在运行的线程或就绪的线程处于这个状态。
（3）BLOCKED：线程因为等待某个被其他线程使用的资源而处于暂停状态。
（4）WAITING：线程因为等待其他线程的数据而处于等待状态。
（5）TIMED_WAITING：线程因为等待其他线程的数据而处于等待状态，只等待指定时间。
（6）TERMINATED：线程结束运行后处于终止状态。

在某个特定的时刻，线程只能处于上述 6 种状态之一。线程的各个状态之间是可以发生迁移的，这种状态迁移是根据线程是否获得了 CPU 的使用权及线程是否执行了某些特定操作而发生的。线程的状态迁移及其发生条件如图 13-2 所示。

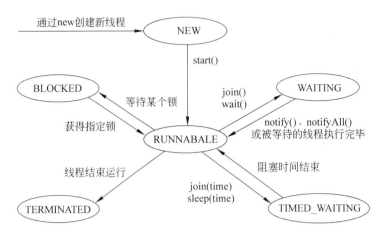

图 13-2 线程的状态迁移及其发生条件

线程状态发生迁移的最终控制者是操作系统中的一个被称为调度器的程序，也就是说，操作系统的调度器程序最终决定把 CPU 分配给哪个线程使用及使用多长时间。不同的操作系统有不同的调度策略，如 Windows 操作系统的调度策略和 Linux 系统的调度策略就不同。由于调用方式和策略超出了本书范围，这里不再赘述。

虽然线程是否被调度的最终控制权在操作系统的调度器及其调度策略手里，但是，可以通过设置线程的优先级参数使线程得到更多被调度器调度的机会。Java 的优先级是一个从 1 到 10 的整数：1 表示最小优先级，10 表示最大优先级，默认优先级是 5。Thread 类中定义了几个优先级常量属性分别表示这些优先级：Thread.MIN_PRIORITY、Thread.MAX_PRIORITY 和 Thread.NORM_PRIORITY。可以通过调用 Thread 类的 setPriority( ) 方法设置线程的优先级。

13.4 线程并发控制

通过多线程编程，将一件大任务分解为多个小任务，并将每个小任务设计为线程，从而提高 CPU 使用效率和整个程序的运行效率。多线程编程在提升效率的同时，也给编程带来了一些挑战。其中最为重要的挑战就是线程的并发。所谓并发，就是当多个线程同时访问一个变量时，保证数据的一致性。为了对线程的并发和同步有一个初步的了解，先看一个简单的例子。

13.4.1 多线程中数据的不一致性现象举例

多线程并发执行时，如果多个线程同时访问同一个变量，可能会导致变量数据的不一致。例如，下面这个例子中，两个线程同时访问一个计数器变量，如果不加以控制，最后会发现计数器变量的值是不正确的。为此，先在 ch13-01 工程下新建一个名为 com.ttt.conflict 的包，在这个包下新建一个名为 Conflictor 的类。Conflictor 类的代码如下（源代码为 13-06.java）：

```java
package com.ttt.conflict;
public class Conflictor {
    private int counter;
    Thread t1, t2;
    public Conflictor() {
        this.counter = 0;
        t1 = new Thread(new MyRunnable(), "第一个线程");
        t2 = new Thread(new MyRunnable(), "第二个线程");
        t1.start();
        t2.start();
    }
    public void waiting() {
        try {
            t1.join();
            t2.join();
        } catch (InterruptedException e) {
            throw new RuntimeException(e);
        }
    }
    public void display() {
        System.out.println("counter: " + counter);
    }
    private class MyRunnable implements Runnable {
        @Override
        public void run() {
            int i = 0;
            while(i<10000) {
                counter++;
                i++;
            }
        }
    }
}
```

在 Conflictor 的构造方法中，初始化 counter 的值为 0，创建了两个名字分别为"第一个线程"和"第二个线程"的线程并启动它们运行。

```java
public Conflictor() {
    this.counter = 0;
    t1 = new Thread(new MyRunnable(), "第一个线程");
    t2 = new Thread(new MyRunnable(), "第二个线程");
    t1.start();
    t2.start();
}
```

其中的 MyRunnable 是实现了 Runnable 接口的类。

```
private class MyRunnable implements Runnable {
    @Override
    public void run() {
        int i = 0;
        while(i<10000) {
            counter++;
            i++;
        }
    }
}
```

在其 run( ) 方法中,通过一个循环将 counter 变量的值每次加 1。

因为两个线程都对 counter 变量进行了 10000 次的加 1 操作,正常情况下,最终 counter 变量的值应该是 20000。现在修改 Main 类,创建 Conflitor 类的对象,在 main 线程中等待 Conflictor 的两个线程运行结束,然后显示 counter 的值。修改后的 Main 类代码如下(源代码为 13-07.java):

```
package org.example;
import com.ttt.conflict.Conflictor;
public class Main {
    public static void main(String[] args) {
        Conflictor c = new Conflictor();
        c.waiting();
        c.display();
    }
}
```

运行这个程序,显示以下信息。

```
counter: 16948
```

再次运行这个程序,又显示以下信息。

```
counter: 17351
```

不仅运行结果不正确,而且每次运行的结果还不一样。为什么会这样呢?分析这个出现这个结果的原因在于:因为两个线程在同时运行,且 CPU 的速度非常快,可能在其中一个线程为 counter 变量执行加 1 操作时,另一个线程又对 counter 执行了加 1 操作,从而覆盖掉了前一个线程的数据。这种现象在多线程程序中经常出现。那么,如何来解决这个问题呢?有多种解决这个问题的办法。先看第一种,也是非常简单的一种解决办法:使用 synchronized 关键字。

13.4.2　使用 synchronized 控制线程并发

这是非常简单的解决多线程并发的方式：对于被多个线程访问的数据，为共享数据，使用 synchronized 关键字对数据访问进行控制。Java 的 synchronized 关键字的作用是：当一个线程执行被 synchronized 关键字修饰的代码块时，如果此时其他线程也需要执行这些代码块，则其他线程需要等待，直到前一个线程执行完这个代码块。

现在，使用 synchronized 关键字来修正第 13.4.1 小节出现数据错误的例子。经过分析发现，程序之所以出现数据不一致问题，是因为两个线程都需要访问 counter 变量。因此，可以使用 synchronized 关键字对 counter 的访问进行控制。修改后的 Conflictor 代码如下（源代码为 13-08.java）：

```java
package com.ttt.conflict;
public class Conflictor {
    private int counter;
    Thread t1, t2;
    public Conflictor() {
        this.counter = 0;
        t1 = new Thread(new MyRunnable(), "第一个线程");
        t2 = new Thread(new MyRunnable(), "第二个线程");
        t1.start();
        t2.start();
    }
    public void waiting() {
        try {
            t1.join();
            t2.join();
        } catch (InterruptedException e) {
            throw new RuntimeException(e);
        }
    }
    public void display() {
        System.out.println("counter: " + counter);
    }
    synchronized private void addOneToCounter() {
        counter++;
    }
    private class MyRunnable implements Runnable {
        @Override
        public void run() {
            int i = 0;
            while (i < 10000) {
                //counter++;
                addOneToCounter();
```

```
                i++;
            }
        }
    }
}
```

在新修改的 Conflictor 类的代码中，使用方法 addOneToCounter( ) 对 counter 变量进行操作，并使用 synchronized 关键字对这个方法加以修饰。同时，将原来在线程中直接对 counter 的加 1 操作修改为对方法 addOneToCounter( ) 的调用。这样，就形成了对 counter 变量的并发控制。Main 类不需要做任何修改。现在运行这个程序，显示以下结果。

```
counter: 20000
```

结果是完全正确的。

synchronized 关键字不仅可以修饰一个方法，还可以对一个代码片段进行修饰，以保证所修饰的代码片段在一个时刻只能被一个线程执行。现在，再次修改 Conflictor 类，本次使用 synchronized 关键字修饰一个需要保护的代码片段。修改后的 Conflictor 类代码如下（源代码为 13-09.java）：

```
package com.ttt.conflict;
public class Conflictor {
    private final static Object lock = new Object();
    private int counter;
    Thread t1, t2;
    public Conflictor() {
        this.counter = 0;
        t1 = new Thread(new MyRunnable(), "第一个线程");
        t2 = new Thread(new MyRunnable(), "第二个线程");
        t1.start();
        t2.start();
    }
    public void waiting() {
        try {
            t1.join();
            t2.join();
        } catch (InterruptedException e) {
            throw new RuntimeException(e);
        }
    }
    public void display() {
        System.out.println("counter: " + counter);
    }
    private class MyRunnable implements Runnable {
```

```
        @Override
        public void run() {
            int i = 0;
            while (i < 10000) {
                synchronized(lock) {
                    counter++;
                }
                i++;
            }
        }
    }
}
```

在修改的 Conflictor 类中，定义了一个普通的 Object 类型变量。

```
private final static Object lock = new Object();
```

这个变量经常被称为监视器变量，这是在 synchronized 关键字监视代码片段时需要使用的一个参数。注意，这个参数不一定必须是 Object 类型的对象，原则上可以是任意类型的对象。然后，在 MyRunnable 类中，将对 counter 变量的加 1 操作修改如下：

```
synchronized(lock) {
    counter++;
}
```

也就是说，通过 synchronized 关键字结合监视器对象保证锁修饰的代码在一个时刻只有一个线程在执行，从而实现了线程的并发控制。执行这个程序，会得到完全正确的结果。

13.4.3 使用原子类型变量控制线程并发

对于像 int 类型整数、long 类型整数和 boolean 类型的布尔变量，Java 提供了称为原子数据类型的封装类型来完成对这些数据的原子操作。所谓原子操作，就是在一个线程操作这种类型的变量时，如果其他线程需要访问同一个变量则必须等待。现在修改 Conflictor 类，使用原子数据类型变量保存 counter 的值。修改后的 Conflictor 类代码如下（源代码为 13-10.java）：

```
package com.ttt.conflict;
import java.util.concurrent.atomic.AtomicInteger;
public class Conflictor {
    private AtomicInteger counter;
    Thread t1, t2;
    public Conflictor() {
```

```java
            counter = new AtomicInteger(0);
            t1 = new Thread(new MyRunnable(), "第一个线程");
            t2 = new Thread(new MyRunnable(), "第二个线程");
            t1.start();
            t2.start();
        }
        public void waiting() {
            try {
                t1.join();
                t2.join();
            } catch (InterruptedException e) {
                throw new RuntimeException(e);
            }
        }
        public void display() {
            System.out.println("counter: " + counter.get());
        }
        private class MyRunnable implements Runnable {
            @Override
            public void run() {
                int i = 0;
                while (i < 10000) {
                    //counter++;
                    counter.addAndGet(1);
                    i++;
                }
            }
        }
    }
```

在新修改的 Conflictor 类中，将原来的 int 类的 counter 属性修改为 AtomicInteger 类型。

```
private AtomicInteger counter;
```

并在构造方法中创建 AtomicInteger 对象，初始化其值为 0。然后在 MyRunnable 类中，将 counter 的加 1 操作修改为对 counter 对象的方法调用，同样实现加 1 功能。

```
counter.addAndGet(1);
```

最后，使用语句：

```
System.out.println("counter: " + counter.get());
```

再显示 counter 的值。运行这个程序，结果是完全正确的。

AtomicInteger 类的常用方法如表 13-3 所示。JDK 中类似的类还有 AtomicLong 类、AtomicBoolean 类、AtomicIntegerArray 类等。

表 13-3 AtomicInteger 类的常用方法

方　　法	功　　能
AtomicInteger( )	构造方法。创建 AtomicInteger 对象并初始化其值为 0
AtomicInteger(int initialValue)	构造方法。创建 AtomicInteger 对象并初始化其值为指定的参数值
final int addAndGet(int delta)	使对象的值加上 delta，并返回结果值
final int decrementAndGet( )	使对象的值减去 delta，并返回结果值
final int get( )	返回对象的 int 值
final int getAndAdd(int delta)	使对象的值加上 delta，并返回以前的值
final int getAndSet(int newValue)	设置对象的值为参数 newValue 的值，并返回以前的值
final void set(int newValue)	设置对象的值为参数 newValue 的值

13.4.4　使用 Lock 接口控制线程并发

Java 中非常灵活的线程并发控制机制是使用 Lock 接口及其实现类提供的多种方式。这些接口及其实现类定义在 Java 的 java.util.concurrent.locks 包中。限于篇幅，本书不对 Lock 的使用进行介绍。

13.5　线程同步控制及生产者—消费者模型

所谓线程同步，是指当多个线程协同完成一项任务时，保证线程之间能够协调一致地运行。线程同步控制的典型应用是生产者—消费者模型：一个线程负责产生数据，另一个线程负责处理数据。这两个线程必须协调一致地工作：消费者线程必须等待生产者产生数据后才可以进行处理；而生产者必须等到消费者处理完已有数据之后才可以继续提供数据。Java 为生产者—消费者模型提供了良好的支持：使用 synchronized 关键字和所有对象从 Object 类继承的 wait( ) 方法和 notify( ) 方法。下面举例说明如何实现线程的同步。

线程同步控制及生产者—消费者模型

13.6　线　程　池

影响 Java 程序执行效率的最大因素是频繁地创建和销毁对象，因为创建一个对象需要为对象分配内存和初始化内存。因此，为了提升 Java 程序的运行效率，提出了线程池和线程工厂模型。所谓线程池，就是预先创建好的一组线程对象，在需要执行某个任务时，可以从线程池中取得一个线程对象，并委托线程对象执行指定的任务。Java 线程都是 Thread 类的对象，因此，通过使用线程池和线程工厂可以有效提升 Java 程序的执行效率。

13.6.1 Java 线程池框架

在使用线程池之前，需要先获得一个线程池对象。Java 线程池框架提供了 Executors 工具类，调用这个工具类的静态方法可以获取不同类型的线程池对象。Executors 类的常用方法如表 13-4 所示。

表 13-4 Executors 类的常用方法

方 法	功 能
static ExecutorService newFixedThreadPool(int nThreads)	创建线程个数由参数 nThreads 指定的线程池。当线程池执行的线程数大于 nThreads 时，后续线程必须等待
static ExecutorService newFixedThreadPool(int nThreads, ThreadFactory threadFactory)	创建线程个数由参数 nThreads 指定的线程池，并且，采用指定的线程工厂参数 threadFactory 创建新线程。当线程池执行的线程数大于 nThreads 时，后续线程必须等待
static ExecutorService newSingleThreadExecutor( )	创建只有一个线程的线程池，所有提交给这类线程池的线程只能顺序被执行
static ScheduledExecutorService newSingleThreadScheduledExecutor( )	创建只有一个线程并且线程可以被延时调度的线程池，所有提交给这类线程池的线程只能顺序被执行
static ExecutorService newCachedThreadPool( )	创建一个无数量限制的线程池
static ScheduledExecutorService newScheduledThreadPool(int corePoolSize)	创建具有指定数量线程对象并且可延时调度的线程池

如表 13-4 所示，使用 Executors 类的静态方法创建线程池的返回结果类型有两个：ExecutorService 接口和 ScheduledExecutorService 接口。并且，ScheduledExecutorService 接口是 ExecutorService 接口的子接口。它们的常用方法分别如表 13-5 和表 13-6 所示。

表 13-5 ExecutorService 接口的常用方法

方 法	功 能
Future<?> submit(Runnable task)	使用一个空闲线程执行由参数 task 指定的任务
<T> Future<T> submit(Runnable task, T result)	使用一个空闲线程执行由参数 task 指定的任务，线程结束后返回 result 结果
<T> Future<T> submit(Callable<T> task)	使用一个空闲线程执行由参数 task 指定的任务，并返回执行结果
void shutdown( )	关闭线程池，继续执行旧任务直到结束，不再接受新任务
boolean isShutdown( )	判断一个线程池是否被关闭
boolean isTerminated( )	判断线程池被关闭后，是否所有线程都已执行完毕

表 13-6 ScheduledExecutorService 接口的常用方法

方 法	功 能
ScheduledFuture<?>schedule(Runnable command, long delay, TimeUnit unit)	按指定的延时参数 delay 和 unit 执行由 command 参数指定的任务

续表

方法	功 能
<V> ScheduledFuture<V>schedule(Callable<V> callable, long delay, TimeUnit unit)	按指定的延时参数 deley 和 unit 执行由 callable 参数指定的任务
ScheduledFuture<?>scheduleAtFixedRate(Runnable command, long initialDelay, long period, TimeUnit unit)	按指定的初始延时和周期执行由 command 指定的任务
ScheduledFuture<?>scheduleWithFixedDelay(Runnable command, long initialDelay, long delay, TimeUnit unit)	按指定的初始延时和间隔执行由 command 指定的任务

注意：ScheduledExecutorService 接口是 ExecutorService 接口的子接口，所以，它具有如表 13-5 所示的所有方法。

13.6.2 线程池使用举例

这个例子首先创建一个名为 MePrinterCallable 的类，这个类实现了 Callable 接口。在它的 call( ) 方法中打印自身线程名称的线程，休眠 1 秒，然后返回一个随机整数。MePrinterCallable 类的代码如下（源代码为 13-11.java）：

```java
package com.ttt.pool;
import java.util.Random;
import java.util.concurrent.Callable;
import java.util.concurrent.atomic.AtomicInteger;
public class MePrinterCallable implements Callable<Integer> {
    private static final AtomicInteger task_no = new AtomicInteger(0);
    private final String task_name;
    private static final Random random = new Random(System.currentTimeMillis());;
    public MePrinterCallable() {
        task_name = "Task-" + task_no.incrementAndGet();
    }
    @Override
    public Integer call() {
        System.out.println("线程: '" + task_name + "' 在运行。");
        try {
            Thread.sleep(1000);
        } catch (InterruptedException e) {
            throw new RuntimeException(e);
        }
        return random.nextInt(1000, 2000);
    }
}
```

为了给每个将 MePrinterCallable 类的对象作为任务运行的线程唯一编号，定义了一个原子类型的属性变量。

```
private static final AtomicInteger task_no = new AtomicInteger(0);
```

解决多线程访问 task_no 变量时的并发控制。同时，在构造方法中使用以下语句。

```
task_name = "Task-" + task_no.incrementAndGet();
```

为了使每个线程具有唯一的名字，在 MePrinterCallable 类的 call( ) 方法中，通过以下语句：

```
System.out.println("线程: '" + task_name + "' 在运行。");
```

打印线程的名字，休眠 1 秒，最后返回一个 int 类型的随机数。

现在修改 Main 类的代码为以下代码（源代码为 13-12.java）。

```java
package org.example;
import com.ttt.pool.MePrinterCallable;
import java.util.concurrent.ExecutionException;
import java.util.concurrent.ExecutorService;
import java.util.concurrent.Executors;
import java.util.concurrent.Future;
public class Main {
    public static void main(String[] args) {
        ExecutorService pool = Executors.newFixedThreadPool(3);
        Future<?>[] fs = new Future[7];
        for(int i=0; i<7; i++) {
            MePrinterCallable epc = new MePrinterCallable();
            fs[i] = pool.submit(epc);
        }
        for (Future<?> f : fs) {
            try {
                Integer o = (Integer)f.get();
                System.out.println(o);
            } catch (InterruptedException | ExecutionException e) {
                throw new RuntimeException(e);
            }
        }
        pool.shutdown();
    }
}
```

在 main( ) 方法中，首先使用语句：

```
ExecutorService pool = Executors.newFixedThreadPool(3);
```

创建了有三个线程的线程池。为了保存线程运行结束后的返回值,定义了 Future 类型的数组。

```
Future<?>[] fs = new Future[7];
```

因为后续会创建 7 个线程,所以这里的数组大小为 7。之后,使用语句:

```
MePrinterCallable epc = new MePrinterCallable();
fs[i] = pool.submit(epc);
```

创建了任务对象并将任务对象提交给线程池执行,同时将任务的执行结果保存在 fs 相应元素中。之后,通过以下循环语句:

```
for (Future<?> f : fs) {
    try {
        Integer o = (Integer)f.get();
        System.out.println(o);
    } catch (InterruptedException | ExecutionException e) {
        throw new RuntimeException(e);
    }
}
```

显示线程的执行结果。现在,运行这个程序,显示以下结果。

```
线程:'Task-1'在运行。
线程:'Task-2'在运行。
线程:'Task-3'在运行。
线程:'Task-4'在运行。
1929
1127
线程:'Task-5'在运行。
1127
线程:'Task-6'在运行。
1127
线程:'Task-7'在运行。
1127
1127
1127
```

从运行结果可以看出,在显示:

```
线程:'Task-1'在运行。
线程:'Task-2'在运行。
线程:'Task-3'在运行。
```

这段信息后会有一个 1 秒的简短暂停，这说明，线程池已被三个线程全部占用，没有新的线程可以执行新的任务，后续任务只能等待，直到线程池中有空闲的线程后再执行新的任务。当任务执行完成后，程序会显示各个任务的返回值。由于有些线程已经执行完毕，而有的线程还在执行，所以出现了显示返回值与显示线程名字的信息交替出现的现象。

如果将线程池的大小从 3 改为大于 7 的任何一个数，如 30，因为线程池中有足够的线程对象可以执行任务，所以，显示的结果将这样变化：先显示线程运行信息，再显示线程返回值信息。

13.6.3　多例多线程和单例多线程及 ThreadLocal 类的使用

在第 13.6.2 小节的例子中，Main 类在创建线程池之后，创建线程任务并委托线程池的线程执行任务的代码，如下加黑代码所示。

```
ExecutorService pool = Executors.newFixedThreadPool(3);
Future<?>[] fs = new Future[7];
for(int i=0; i<7; i++) {
    MePrinterCallable epc = new MePrinterCallable();
    fs[i] = pool.submit(epc);
}
```

在 for 循环中，每次都创建了一个 MePrinterCallable 类的对象，并通过线程池的 submit( ) 方法将这个对象所指代的任务委托给线程池中的线程执行。在这里，线程池中的每个线程所执行的任务是由不同的对象所指代的：7 个任务由 7 个 MePrinterCallable 类的对象指代，并委托给 7 个线程执行，这种方式称为多例多线程。当然，也可以将以上代码改成如下所示的形式。

```
ExecutorService pool = Executors.newFixedThreadPool(3);
Future<?>[] fs = new Future[7];
MePrinterCallable epc = new MePrinterCallable();
for(int i=0; i<7; i++) {
    fs[i] = pool.submit(epc);
}
```

只是将语句：

```
MePrinterCallable epc = new MePrinterCallable();
```

提到了 for 循环的外面，但是本质上却发生了很大的变化：7 个线程执行由 1 个 MePrinterCallable 类的对象所指代的任务，这种方式称为单例多线程。现在运行这个程序，显示以下结果。

```
线程: 'Task-1' 在运行。
线程: 'Task-1' 在运行。
线程: 'Task-1' 在运行。
线程: 'Task-1' 在运行。
线程: 'Task-1' 在运行。
1636
1866
1044
线程: 'Task-1' 在运行。
1151
1410
1892
1296
```

从运行结果看，任务的名字是不对的。为什么会出现这种错误呢？分析 MePrinterCallable 类的代码可以发现，在 MePrinterCallable 类的构造方法中就已经为代表任务名字的 task_name 属性变量赋值了。

```
public MePrinterCallable() {
    task_name = "Task-" + task_no.incrementAndGet();
}
```

导致 7 个线程任务的名字都是 Task-1。如何修改代码使其在单例多线程下可以正确运行呢？可以使用 ThreadLocal 类。

ThreadLocal 类是 Java 提供的一个工具类，这个类可以保证对于定义为 ThreadLocal 类的属性变量，在进行多线程操作时，每个线程都具有自己的属性变量，并且这个属性变量的值在各个线程之间不共享、不冲突。ThreadLocal 类的常用方法如表 13-7 所示。

表 13-7 ThreadLocal 类的常用方法

方法	功能
ThreadLocal( )	构造方法。ThreadLocal 类是泛型类，在定义这个类的变量时需要给出泛型参数 T
void set(T value)	设置 ThreadLocal 对象的值
T get( )	获取 ThreadLocal 对象的值
void remove( )	移除 ThreadLocal 对象的值

现在，修改 MePrinterCallable 类的代码，使之可以在单例多线程模式下作为线程任务被正确执行。修改后的 MePrinterCallable 类代码如下（源代码为 13-13.java）：

```java
package com.ttt.pool;
import java.util.Random;
import java.util.concurrent.Callable;
import java.util.concurrent.atomic.AtomicInteger;
public class MePrinterCallable implements Callable<Integer> {
    private static final AtomicInteger task_no = new AtomicInteger(0);
    private final ThreadLocal<String> task_name;
    private static final Random random = new Random(System.current
    TimeMillis());;
    public MePrinterCallable() {
        task_name = new ThreadLocal<>();
    }
    @Override
    public Integer call() {
        task_name.set("Task-" + task_no.incrementAndGet());
        System.out.println("线程：'" + task_name.get() + "' 在运行。");
        try {
            Thread.sleep(1000);
        } catch (InterruptedException e) {
            throw new RuntimeException(e);
        }
        return random.nextInt(1000, 2000);
    }
}
```

程序首先定义了一个 ThreadLocal 类的属性变量 task_name。

```
private final ThreadLocal<String> task_name;
```

因为是 ThreadLocal 类的变量，因此可以保证当同一个 MePrinterCallable 类的对象所指代的任务被多个线程执行时，每个线程的 task_name 属性变量的值不共享、不冲突。简单来说就是：每个线程都有自己的 task_name 属性变量，然后在 MePrinterCallable 类的构造方法中创建了 task_name 的对象。

```
task_name = new ThreadLocal<>();
```

进而在 call( ) 方法中设置 task_name 变量为各自不同的值，并显示线程名称信息。

```
task_name.set("Task-" + task_no.incrementAndGet());
System.out.println("线程：'" + task_name.get() + "' 在运行。");
```

现在运行修改后的 MePrinterCallable 类和 Main 类，显示以下信息。

```
线程:'Task-2'在运行。
线程:'Task-3'在运行。
线程:'Task-1'在运行。
线程:'Task-4'在运行。
线程:'Task-5'在运行。
1061
1842
线程:'Task-6'在运行。
1287
线程:'Task-7'在运行。
1907
1777
1999
1725
```

结果完全正确。

多例多线程和单例多线程是程序设计中常用的设计方式,但是,就应用情况看,由于单例多线程更节省资源(因为只需要一个指代任务的对象),所以,应用更广泛。例如,书名的 Spring 框架默认情况下使用的就是单例多线程方式;著名的容器服务器 Tomcat 使用的也是单例多线程方式。

在单例多线程方式中,需要仔细分析哪些属性变量需要隔离,也就是不同的线程需要有各自隔离的属性变量;哪些属性需要共享,也就是不同的线程需要共享属性变量的值。这些问题与业务相关,但没有标准解决方案。多线程编程一直处于程序设计难度排行的前列。

13.7 Java 线程应用实践

Java 提供了多种创建线程的方式和解决线程数据并发的方式,那么,在编程实践中应该如何选用呢?

对于创建线程,应用实践是:如果只需要创建一个完成某个任务的线程,并且不关心线程的返回值,那么,使用通过实现 Runnable 方式是最佳选择;如果需要创建多个用于完成某个任务的线程,并且不关心线程的返回值,那么,使用通过继承 Thread 类实现自己的子类方式是最佳选择,因为通过 new 这个子类,即可以创建多个同类型的线程;如果要从线程中获取线程的运行结果值,只能采用 FutureTask 方式。

对于解决线程的数据并发,应用实践是:除非需要对并发进行灵活的控制,否则,使用 synchronized 关键字永远是最佳选择。其次,如果程序需要控制并发的数据是一些基本数据类型,如 int 类型、long 类型,那么,使用原子数据类型控制并发也是一个不错的选择。

对于线程池,除非在程序中需要使用大量线程完成任务,否则,在现阶段使用这种技术进行编程实践的机会不大。线程池是后续很多内容的支撑点,如 Java Web 编程,因此,了解了线程池技术以后,可以在后续课程中做到知其所以然。

总之，请记住，在程序设计世界里，简单的就是最好的。因为简单，所以不易出错；因为"简单"，即使出现错误也容易定位和解决。不要在程序设计中炫技，更不要把简单问题复杂化。

13.8 案例：找出小于 1000 的所有质数和水仙花数

所谓质数，就是只能被 1 和自己整除的整数。

13.8.1 案例任务

设计两个线程，分别用于找出小于 1000 的所有质数和小于 1000 的所有水仙花数，然后，主线程等待这两个线程运行结束后，在主线程中显示找到的所有结果。

13.8.2 任务分析

程序要求在两个线程中分别找到质数和水仙花数，并在主线程中显示满足要求的质数和水仙花数，因此，应该采用有返回值的方式创建线程，也就是采用 FutureTask 方式创建这两个线程。程序可以设计以下几个类完成相关任务。

（1）PrimeFinder 类：寻找 1000 以内的质数，并返回结果给调用者。
（2）NarcissisticFinder 类：寻找 1000 以内的水仙花数，并返回结果给调用者。
（3）Main 类：测试类。

13.8.3 任务实施

在 ch13-01 工程下新建一个名为 com.ttt.task 的包，在这个包下新建 PrimeFinder 类和 NarcissisticFinder 类。PrimeFinder 类的代码如下（源代码为 13-14.java）：

```java
package com.ttt.task;
import java.util.ArrayList;
import java.util.List;
import java.util.concurrent.Callable;
import java.util.concurrent.ExecutionException;
import java.util.concurrent.FutureTask;
public class PrimeFinder {
    private List<Integer> result;
    private FutureTask<List<Integer>> ft;
    public PrimeFinder() {
        result = new ArrayList<>();
        ft = new FutureTask<>(new Callable<List<Integer>>() {
```

```
            @Override
            public List<Integer> call() throws Exception {
                for(int i=2; i<1000; i++) {
                    if (isPrime(i)) result.add(i);
                }
                return result;
            }
        });
    }
    public List<Integer> go() {
        Thread t = new Thread(ft);
        t.start();
        try {
            return ft.get();
        } catch (InterruptedException | ExecutionException e) {
            throw new RuntimeException(e);
        }
    }
    private boolean isPrime(int n) {
        int m = n/2;
        for(int i=2; i<=m; i++)
            if (n%i == 0) return false;
        return true;
    }
}
```

NarcissisticFinder 类的代码如下（源代码为 13-15.java）：

```
package com.ttt.task;
import java.util.ArrayList;
import java.util.List;
import java.util.concurrent.Callable;
import java.util.concurrent.ExecutionException;
import java.util.concurrent.FutureTask;
public class NarcissisticFinder {
    private List<Integer> result;
    private FutureTask<List<Integer>> ft;
    public NarcissisticFinder() {
        result = new ArrayList<>();
        ft = new FutureTask<>(() -> {
            for(int i=1; i<1000; i++) {
                if (isNarcissistic(i)) result.add(i);
            }
            return result;
        });
```

```
        }
    public List<Integer> go() {
        Thread t = new Thread(ft);
        t.start();
        try {
            return ft.get();
        } catch (InterruptedException | ExecutionException e) {
            throw new RuntimeException(e);
        }
    }
    private boolean isNarcissistic(int d) {
        String sd = "" + d;
        String[] digits = sd.split("");
        int dd = 0;
        for (String digit : digits) {
            dd += Math.pow(Integer.parseInt(digit), digits.length);
        }
        return dd == d;
    }
}
```

Main 类代码如下（源代码为 13-16.java）：

```
package org.example;
import com.ttt.task.NarcissisticFinder;
import com.ttt.task.PrimeFinder;
import java.util.List;
public class Main {
    public static void main(String[] args) {
        PrimeFinder pf = new PrimeFinder();
        List<Integer> p = pf.go();
        NarcissisticFinder nf = new NarcissisticFinder();
        List<Integer> n = nf.go();
        System.out.println("1000 以内所有质数：");
        p.forEach(System.out::println);
        System.out.println("1000 以内所有水仙花数：");
        n.forEach(System.out::println);
    }
}
```

运行这个程序，显示以下结果。

```
1000 以内所有质数：
2
```

```
...// 省略了部分信息
997
1000 以内所有水仙花数：
1
...// 省略了部分信息
371
407
```

13.9 练习：统计上网人数

一个网站的访问量反映了一个网站的活跃程度。使用一个全局变量 count 记录当前在线的网民人数。设计一个线程模拟上网行为：先对 count 变量加 1，模拟一个新的网民正在在线，然后随机休眠一段时间，再将 count 变量减 1，模拟一个网民下线行为。在主线程中，创建 1000 个这样的线程对 count 变量进行操作，并实时显示 count 的值以表示当前在线人数。

第 14 章 网 络 编 程

现代应用程序都不可避免地要与外界通信。例如，在 QQ 聊天程序中，QQ 用户向好友发送信息，QQ 用户接收好友发来的信息；在使用浏览器访问网络信息时，浏览器从指定的服务器获取页面信息，通过浏览器解析后显示在浏览器中。因此，Java 程序实现网络通信功能是一项非常重要的能力。本章对使用 Java 实现网络编程进行介绍。

14.1 网络通信协议

网络上任何两台计算机之间要进行通信，必须按照一种事先约定好的规则进行，这种规则称为通信协议。常见的通信协议有很多种，如 x.25、DDN、TCP/IP 等。目前使用的非常多也非常广泛的协议是 TCP/IP，它是 Internet（因特网）的基础：Internet 网上的任何计算机要与其他计算机通信，必须使用 TCP/IP 进行。TCP/IP 的核心内容是 IP 地址和端口。下面就两个问题进行介绍。

IP 地址和端口的使用情况观察

14.1.1 IP 地址 InetAddress 类和端口

在基于 TCP/IP 的网络中，每台计算机都会被分配一个唯一的地址，这个地址称为 IP 地址。TCP/IP 目前有两个版本：IPv4 版本和 IPv6 版本。它们处理通信过程的方式是类似的，不同之处在于地址的长度不同。IPv4 版本的 IP 地址是用由"."分隔的 4 个 0~255 的整数表示；而 IPv6 的地址则是用由"."分隔的 16 个 0~255 的整数表示。例如，122.10.0.44 就是一个合法的 IPv4 地址；255.255.255.255 也是一个合法的 IPv4 地址。

在 JDK 中专门定义了 InetAddress 类来表示一个 IP 地址。InetAddress 类的常用方法如表 14-1 所示。

表 14-1 InetAddress 类的常用方法

方　　法	功　　能
static InetAddress getByAddress(byte[] addr)	基于给定的数组参数构造 InetAddress 对象
static InetAddress getByName(String host)	基于给定的名字参数构造 InetAddress 对象
static InetAddress getLocalHost( )	获取本机的 IP 地址
String getHostAddress( )	获取字符串形式的 IP 地址
String getHostName( )	获取主机的名字
boolean isReachable(int timeout)	判断是否可以与指定的地址计算机进行通信

通常来说，现代计算机一般会同时运行多个需要与外界进行网络通信的程序。例如，在计算机上使用浏览器浏览网页的同时，还可能运行了 QQ 聊天程序等。对于多个在同一台计算机上运行的程序，它们的 IP 地址是一样的，那么，计算机是如何保证不同的程序得到属于自己的从外界发来的信息而不至于造成混淆呢？答案就是端口：给每个程序分配一个唯一的端口。也就是说，计算机的 IP 地址和端口唯一地定位了计算机上运行的一个通信程序，有时也称计算机 IP 地址和端口唯一地确定了通信的端点。因此，在编写通信程序时，为了使信息无差错地发送到另一台计算机的某个程序中，必须指明目标计算机的 IP 地址和程序端口。下面编写一个能够获取本机 IP 地址的简单程序。

为此，在 IDEA 中新建一个名为 ch14-01 的 Java 工程。新建的 ch14-01 工程如图 14-1 所示。

图 14-1　新建的 ch14-01 工程

然后修改 Main 类为以下代码（源代码为 14-01.java）。

```
package org.example;
import java.net.InetAddress;
import java.net.UnknownHostException;
public class Main {
    public static void main(String[] args) {
        try {
            InetAddress addr = InetAddress.getLocalHost();
            System.out.println(addr.getHostName());
            System.out.println(addr.getHostAddress());
        } catch (UnknownHostException e) {
            throw new RuntimeException(e);
        }
    }
}
```

在 main( ) 方法中，使用以下语句：

```
InetAddress addr = InetAddress.getLocalHost();
```

获取了包含本机 IP 地址的 InetAddress 对象，然后使用语句：

```
System.out.println(addr.getHostName());
System.out.println(addr.getHostAddress());
```

显示主机名和 IP 地址。运行这个程序，显示以下信息。

```
DESKTOP-Q5E6PT1
192.168.160.132
```

14.1.2 UDP 和 TCP

TCP/IP，从本质上说是一个通信协议家族，也就是说 TCP/IP 是一组协议的总称。其中包括 ICMP、UDP、TCP、FTP、HTTP 等。UDP 和 TCP 是最为常用的两个协议。下面两节对如何使用它们编写通信程序进行介绍。

14.2 使用 UDP 进行通信

UDP 称为数据包协议，是一种无连接的数据对等通信协议，也就是说通信的双方没有主从关系，任何一方只要知道对方的 IP 地址和端口即可进行通信。Java 对 UDP 进行了抽象：使用 DatagramSocket 类表示双方程序的通信端点；使用 DatagramPacket 类表示要进行通信的数据信息。

14.2.1 DatagramSocket 类和 DatagramPacket 类

Java 使用 DatagramSocket 类表示通信双方的端点，使用 DatagramPacket 类表示要发送或接收的数据：在可以使用 UDP 发送数据消息之前，程序需要创建表示自己端点的 DatagramSocket 对象，然后创建 DatagramPacket 类的对象，在其中设置好要发送的数据、消息目标方的 IP 地址和端口，最后调用 DatagramPacket 对象的 send( ) 方法发送数据到对方；类似地，在可以使用 UDP 接收数据消息之前，程序需要创建表示自己端点的 DatagramSocket 对象，然后创建 DatagramPacket 类的对象，在其中设置好用于存放接收消息的缓冲区，最后调用 DatagramPacket 对象的 receive( ) 方法接收数据消息。DatagramPacket 类的常用方法如表 14-2 所示。

表 14-2 DatagramSocket 类的常用方法

方　　法	功　　能
DatagramSocket( )	构造方法。创建 DatagramSocket 的对象，绑定端点地址为主机的任何可用地址，端口为任何可用端口
DatagramSocket(int port)	构造方法。创建 DatagramSocket 的对象，绑定端点地址为主机的任何可用地址，端口为参数 port 指定的端口

续表

方 法	功 能
DatagramSocket(int port, InetAddress iaddr)	构造方法。创建 DatagramSocket 的对象，绑定端点地址为参数 iaddr 指定的地址，端口为参数 port 指定的端口
void close( )	关闭端点，释放所占用的端口
InetAddress getInetAddress( )	获取这个端点绑定的 InetAddress 对象
int getPort( )	获取这个端点绑定的端口号
void send(DatagramPacket p)	发送 p 中的数据到参数 p 所指定的端点
void receive(DatagramPacket p)	接收其他计算机发来的数据，并保存到参数 p 中

Java 使用 DatagramPacket 类表示要发送的数据和消息目的地的 IP 地址及端口。在发送方可以发送数据之前，需要创建 DatagramPacket 类的对象，在其中设置消息目标方的 IP 地址和端口。DatagramPacket 类的常用方法如表 14-3 所示。

表 14-3　DatagramPacket 类的常用方法

方 法	功 能
DatagramPacket(byte[] buf, int length, InetAddress address, int port)	构造方法。用于创建一个用于发送消息的 DatagramPacket 对象，要发送的数据在 buf 数组中，长度为 length。目标的地址在 address 参数中，端口在 port 参数中
DatagramPacket(byte[] buf, int offset, int length, InetAddress address, int port)	构造方法。用于创建一个用于发送消息的 DatagramPacket 对象，要发送的数据在 buf 数组中的从 offset 开始处，长度为 length。目标的地址在 address 参数中，端口在 port 参数中
DatagramPacket(byte[] buf, int length)	构造方法。用于创建一个用于接收消息的 DatagramPacket 对象，接收到的消息存储于 buf 数组中，数据的长度最长为 length
DatagramPacket(byte[] buf, int offset, int length)	构造方法。用于创建一个用于接收消息的 DatagramPacket 对象，接收到的消息存储于 buf 数组从 offset 开始处，数据消息的长度最长为 length
InetAddress getAddress( )	获取与该数据消息相关联端点的 InetAddress 对象
int getPort( )	获取与该数据消息相关联端点的端口号
byte[] getData( )	获取与该数据消息相关联的数据
int getLength( )	获取与该数据消息相关联的数据长度
void setAddress(InetAddress iaddr)	设置目的地的 IP 地址
void setPort(int iport)	设置目的地的端口号
void setData(byte[] buf)	设置发送数据，或者用于接收数据的存储区

第 14.2.1 小节将通过举例说明如何使用 UDP 进行点对点的数据消息通信。

14.2.2　UDP 点对点通信程序举例

既然是通信，就一定会涉及至少两个程序：一个发送数据消息，另一个接收数据消息，或者反之。在下面这个例子中，一个称为 UDPP2POne，另一个称为 UDPP2PTwo。UDPP2POne 先向 UDPP2PTwo 发送"你好！"信息，然后，UDPP2PTwo 回送"我很好，你呢?"信息。之后,UDPP2POne 再次发送"再见"信息,最后关闭通信。为此,在 ch14-01 工程下新建名为 com.ttt.udpp2p 的包,在这个包下新建 UDPP2POne 类和 UDPP2PTwo 类。UDPP2POne 类的代码如下(源代码为 14-02.java):

```java
package com.ttt.udpp2p;
import java.io.IOException;
import java.net.*;
public class UDPP2POne {
    public static void main(String[] args) {
        UDPP2POne one = new UDPP2POne();
        one.talk();
    }
    public void talk() {
        try(DatagramSocket dgs = new DatagramSocket(7777)) {
            byte[] data = "你好!".getBytes("UTF-8");
            InetAddress addr = InetAddress.getByName("127.0.0.1");
            DatagramPacket dgp1 = new DatagramPacket(data, data.length,
            addr, 8888);
            System.out.println("发送:你好!");
            dgs.send(dgp1);
            byte[] buf = new byte[1024];
            DatagramPacket dgp2 = new DatagramPacket(buf, buf.length);
            dgs.receive(dgp2);
            System.out.println("接收:" + new String(dgp2.getData(),
            0, dgp2.getLength(), "UTF-8"));
            data = "再见".getBytes("UTF-8");
            dgp1 = new DatagramPacket(data, data.length, addr, 8888);
            System.out.println("发送:再见");
            dgs.send(dgp1);
            Thread.sleep(1000);
        } catch (IOException | InterruptedException e) {
            System.out.println("程序错误,异常信息如下:");
            e.printStackTrace();
        }
    }
}
```

在 UDPP2POne 类中,首先使用语句:

```
DatagramSocket dgs = new DatagramSocket(7777)
```

创建表示自己通信端点的 DatagramSocket 对象,并且将这个对象绑定到 7777 端口上。然后使用语句:

```
byte[] data = "你好!".getBytes("UTF-8");
InetAddress addr = InetAddress.getByName("127.0.0.1");
DatagramPacket dgp1 = new DatagramPacket(data, data.length, addr, 8888);
System.out.println("发送: 你好!");
dgs.send(dgp1);
```

创建要发送的数据信息 DatagramPacket。注意，IP 地址 127.0.0.1 表示通信对方程序也运行在本机上，对方的端口为 8888。然后发送消息到 IP 地址为 127.0.0.1、端口为 8888 的程序中。此时，如果对方在接收数据，则应该可以接收到"你好！"的消息。在发送这条消息后，使用语句：

```
byte[] buf = new byte[1024];
DatagramPacket dgp2 = new DatagramPacket(buf, buf.length);
dgs.receive(dgp2);
System.out.println("接收: " + new String(dgp2.getData(),
                                0, dgp2.getLength(), "UTF-8"));
```

接收从对方发来的消息，并显示在屏幕上。之后，再使用语句：

```
data = "再见".getBytes("UTF-8");
dgp1 = new DatagramPacket(data, data.length, addr, 8888);
System.out.println("发送: 再见");
dgs.send(dgp1);
```

发送"再见"消息。在发送了这条消息后，程序休眠 1 秒，其目的是保证对方收到信息后再关闭代表自己的通信端点。

UDPP2PTwo 类的代码如下（源代码为 14-03.java）：

```
package com.ttt.udpp2p;
import java.io.IOException;
import java.net.DatagramPacket;
import java.net.DatagramSocket;
import java.net.InetAddress;
import java.net.SocketException;
public class UDPP2PTwo {
    public static void main(String[] args) {
        UDPP2PTwo two = new UDPP2PTwo();
        two.talk();
    }
    public void talk() {
        try(DatagramSocket dgs = new DatagramSocket(8888)) {
            System.out.println("Waiting...");
```

```
            byte[] buf = new byte[1024];
            DatagramPacket dgp1 = new DatagramPacket(buf, buf.length);
            dgs.receive(dgp1);
            System.out.println("接收: " + new String(dgp1.getData(), 0,
                dgp1.getLength(),"UTF-8"));
            byte[] data = "我很好，你呢?".getBytes("UTF-8");
            InetAddress addr = dgp1.getAddress();
            int port = dgp1.getPort();
            DatagramPacket dgp2 = new DatagramPacket(data, data.length,
                addr, port);
            System.out.println("发送: 我很好，你呢?");
            dgs.send(dgp2);
            dgs.receive(dgp1);
            System.out.println("接收: " + new String(dgp1.getData(), 0,
                dgp1.getLength(), "UTF-8"));
        } catch (IOException e) {
            throw new RuntimeException(e);
        }
    }
}
```

UDPP2PTwo 类的代码与 UDPP2POne 类的代码是类似的，不再赘述。

在使用 UDP 编写网络通信程序时，一定要注意通信双方的协调性：当一方在发送信息时，另一方一定要处于结束信息的状态；否则，数据消息会丢失。因此，网络通信程序一般都是采用多线程形式进行编程设计：专门有一个线程时刻等待对方的数据，这样，确保数据不会因为没有及时接收而丢失。关于网络的多线程编程，会在第 14.6 节进行专门介绍。

现在运行程序，需要注意的是，一定要运行 UDPP2PTwo。启动 UDPP2PTwo 后使之处于等待信息状态，然后启动 UDPP2POne，此时，UDPP2POne 的结果界面显示以下信息。

```
发送: 你好!
接收: 我很好，你呢?
发送: 再见
```

而 UDPP2PTwo 的结果界面显示以下信息。

```
Waiting...
接收: 你好!
发送: 我很好，你呢?
接收: 再见
```

14.3 使用 TCP 进行通信

TCP 称为传输控制协议,是一种面向连接的数据通信协议。通信的双方具有主从关系。在可以使用 TCP 进行数据通信之前,服务器端的程序需要创建 ServerSocket 类的对象并等待客户端的程序主动与之建立连接。在连接建立之后,双方可以使用 Java I/O 流进行数据通信。

建立 TCP 连接的三次握手

14.3.1 客户/服务器模式

综上所述,使用 TCP 进行数据通信的双方具有主从关系:通信的其中一端需要做好准备并等待与另一端建立连接,只有在连接建立后,双方才可以进行数据通信。习惯上,将被动等待建立连接的一方称为服务器端,将主动建立连接的一方称为客户端。这种方式合起来称为客户/服务器模式。

在客户/服务器模式中,服务器端可以同时建立与多个客户端的连接,并通过已经建立的连接同时与多个客户端进行通信。例如,使用浏览器浏览 Internet 网页的应用,就是一个典型的客户/服务器模型:多个人可以同时使用浏览器浏览服务器上的网页。

14.3.2 ServerSocket 类和 Socket 类

Java 使用 ServerSocket 类表示使用 TCP 进行数据通信的服务器端。在可以使用 TCP 进行通信之前,服务器端需要首先创建 ServerSocket 类的对象,并等待客户端的连接;客户端则需要创建 Socket 类的对象,并主动与服务器端建立连接。服务器端收到从客户端发来的建立连接请求后,可以接受连接请求或拒绝连接请求;如果服务器端接受客户端的连接请求,服务器端就会再创建一个专门与这个客户端进行数据通信的 Socket 对象。之后,服务器端和客户端就可以在各自的 Socket 对象上创建 Java I/O 流,通过 I/O 流将数据发送给对方。

ServerSocket 类和 Socket 类在基于 TCP 的通信中起着非常关键的作用。它们的常用方法分别如表 14-4 和表 14-5 所示。

表 14-4 ServerSocket 类的常用方法

方 法	功 能
ServerSocket(int port)	构造方法。使用给定的端口参数 port 创建一个 ServerSocket 对象
ServerSocket(int port, int backlog)	构造方法。使用给定的端口参数 port 创建一个 ServerSocket 对象,并设置允许等待连接的最大客户端的数目为 backlog
ServerSocket(int port, int backlog, InetAddress bindAddr)	构造方法。使用给定的端口参数 port 和 IP 地址参数 bindAddr 创建一个 ServerSocket 对象,并设置允许等待连接的最大客户端的数目为 backlog
void close( )	关闭服务器对象
Socket accept( )	接受客户端的连接请求,并返回可以与这个客户端进行数据通信的套接字对象

表 14-5　Socket 类的常用方法

方　　法	功　　能
Socket( )	构造方法。创建一个客户端套接字对象
Socket(String host, int port)	构造方法。创建一个客户端套接字对象，并建立与由 host 参数和 port 参数指定的服务器的连接
Socket(InetAddress address, int port)	构造方法。创建一个客户端套接字对象，并建立与由 address 参数和 port 参数指定的服务器的连接
void close( )	关闭 Socket 连接
void connect(SocketAddress endpoint)	建立与由参数 endpoint 指定的服务器的连接，连接成功则直接返回，否则抛出异常
void connect(SocketAddress endpoint, int timeout)	建立与由参数 endpoint 指定的服务器的连接，并指定必须在 timeout 毫秒内完成，否则抛出异常
boolean isConnected( )	判断是否已经建立连接
SocketAddress getRemoteSocketAddress( )	获取通信对方的 SocketAddress 地址
int getPort( )	获取通信对方的端口号
InputStream getInputStream( )	获取一个输入流对象，使用该流可以读取对方发来的数据信息
OutputStream getOutputStream( )	获取一个输出流对象，使用该流可以向对方发送数据信息
void setKeepAlive(boolean on)	设置是否间歇性地发送 KEEP ALIVE 连接维持消息

第 14.3.3 小节通过一个例子说明使用 TCP 进行数据通信。

14.3.3　TCP 通信程序举例

这个例子与第 14.2.2 小节的例子功能类似，只是这里使用 TCP 进行通信。在这个例子中，一个称为 TCPServer，它首先等待与客户端的连接，当有客户端请求建立连接时，接受这个连接并创建一个新线程与客户端进行数据通信，之后，服务器端继续等待与客户端的连接；另一个称为 TCPClient，它建立与服务器端的连接，一旦连接成功后，则开始与服务器端进行数据通信。为此，在 ch14-01 工程下新建名为 com.ttt.tcp 的包，在这个包下新建 TCPServer 类和 TCPClient 类。TCPServer 类的代码如下（源代码为 14-04.java）：

```java
package com.ttt.tcp;
import java.io.IOException;
import java.io.InputStream;
import java.io.OutputStream;
import java.net.ServerSocket;
import java.net.Socket;
public class TCPServer {
    public static void main(String[] args) {
        TCPServer ts = new TCPServer();
        ts.go();
    }
    public void go() {
```

```java
        try(ServerSocket ss = new ServerSocket(8888)) {
            while(true) {
                System.out.println("等待与客户端连接...");
                Socket cs = ss.accept();
                newThreadForThisClient(cs);
            }
        }
        catch (IOException e) {
            System.out.println("程序运行错误,错误原因:");
            e.printStackTrace();
        }
    }
    private void newThreadForThisClient(Socket cs) {
        Thread t = new Thread(new Runnable() {
            @Override
            public void run() {
                try(InputStream is = cs.getInputStream();
                    OutputStream os = cs.getOutputStream();) {
                    byte[] buf = new byte[1024];
                    int len = is.read(buf);
                    System.out.println("收到: " + new String(buf, 0, len,
                    "UTF-8"));
                    byte[] b = "我很好,你呢?".getBytes("UTF-8");
                    System.out.println("发送: " + "我很好,你呢?");
                    os.write(b);
                    len = is.read(buf);
                    System.out.println("收到: " + new String(buf, 0, len,
                    "UTF-8"));
                    cs.close();
                }
                catch (IOException e) {
                    System.out.println("程序运行错误,错误原因:");
                    e.printStackTrace();
                }
            }
        });
        t.start();
    }
}
```

在服务器端代码中,使用语句:

```java
try(ServerSocket ss = new ServerSocket(8888)) {
```

创建端口为 8888 的服务器端套接字对象,然后使用语句:

```
Socket cs = ss.accept();
```

等待与客户端连接。一旦客户端请求连接,使用语句:

```
newThreadForThisClient(cs);
```

创建新的线程具体处理与客户端的数据通信。在 newThreadForThisClient( ) 方法中,使用语句:

```
try(InputStream is = cs.getInputStream();
    OutputStream os = cs.getOutputStream();) {
```

获取与客户端进行通信的 I/O 流对象,然后使用语句:

```
byte[] buf = new byte[1024];
int len = is.read(buf);
System.out.println("收到: " + new String(buf, 0, len, "UTF-8"));
```

接收从客户端发来的信息并显示。再使用语句:

```
byte[] b = "我很好,你呢?".getBytes("UTF-8");
System.out.println("发送: " + "我很好,你呢?");
os.write(b);
```

向客户端发送数据。并再次使用语句:

```
len = is.read(buf);
System.out.println("收到: " + new String(buf, 0, len, "UTF-8"));
cs.close();
```

接收从客户端发来的数据并显示,最后关闭 Socket 套接字。

TCPClient 类的代码如下(源代码为 14-05.java):

```
package com.ttt.tcp;
import java.io.IOException;
import java.io.InputStream;
import java.io.OutputStream;
import java.net.InetSocketAddress;
import java.net.Socket;
public class TCPClient {
    public static void main(String[] args) {
        TCPClient tc = new TCPClient();
```

```java
            tc.go();
    }
    public void go() {
        try(Socket sct = new Socket()) {
            sct.connect(new InetSocketAddress("127.0.0.1", 8888));
            OutputStream os = sct.getOutputStream();
            InputStream is = sct.getInputStream();
            byte[] b = "你好!".getBytes("UTF-8");
            System.out.println("发送: " + "你好!");
            os.write(b, 0, b.length);
            byte[] buf = new byte[1024];
            int len = is.read(buf);
            System.out.println("收到: " + new String(buf, 0, len, "UTF-8"));
            b = "再见".getBytes("UTF-8");
            System.out.println("发送: " + "再见");
            os.write(b, 0, b.length);
            is.close();
            os.close();
        } catch (IOException e) {
            System.out.println("不能连接服务器,错误原因:");
            e.printStackTrace();
        }
    }
}
```

该代码首先使用语句:

```java
try(Socket sct = new Socket()) {
    sct.connect(new InetSocketAddress("127.0.0.1", 8888));
    OutputStream os = sct.getOutputStream();
    InputStream is = sct.getInputStream();
```

创建客户端套接字,建立与服务器端的连接,并获取用于数据通信的 I/O 流对象。之后使用语句:

```java
byte[] b = "你好!".getBytes("UTF-8");
System.out.println("发送: " + "你好!");
os.write(b, 0, b.length);
```

向服务器端发送数据。然后使用语句:

```java
byte[] buf = new byte[1024];
int len = is.read(buf);
```

```
System.out.println("收到: " + new String(buf, 0, len, "UTF-8"));
```

接收从服务器端发来的信息并显示。再使用语句:

```
b = "再见".getBytes("UTF-8");
System.out.println("发送: " + "再见");
os.write(b, 0, b.length);
```

向服务器发送数据。最后,关闭 I/O 流。

运行这个程序,先启动服务器程序,等待客户端连接,再启动客户端代码。此时,服务器端显示以下信息。

```
等待与客户端连接...
等待与客户端连接...
收到: 你好!
发送: 我很好,你呢?
收到: 再见
```

客户端界面显示以下信息。

```
发送: 你好!
收到: 我很好,你呢?
发送: 再见
```

由于服务器端一直在运行,所以会多次启动客户端程序,或者,在多个其他计算机运行这个客户端程序。

14.4 使用 HTTP 访问网络页面

人们在使用浏览器访问网络页面时,浏览器程序与服务器端的 Web 服务器程序进行通信时使用的就是 HTTP。HTTP 是非常常用的通信协议,它构建在 TCP 基础之上,定义了浏览器与 Web 服务器之间进行通信的完整规则。从 Java 11 版本开始,Java 提供了自己的用于访问 Web 服务器的 HTTP 客户端的实现类,这个类就是 HttpClient。

14.4.1 Java 对 HTTP 的实现概述

HTTP 是一种简单的"一问一答"式的应用层通信协议。"一问"就是客户端发送给服务器端的请求;"一答"就是服务器端在处理请求后返回给客户端的结果。Java 采用几个核心类和接口对 HTTP 的实现进行了封装,它们是: HttpClient 类,用于对 HTTP 的客户端进行封装；HttpRequest 类，用于对从客户端发送给服务器端的请求数据进行封装；HttpResponse<T> 接口，用于对从服务器端返回给客户端的响应结果进行封装。HttpClient 类、HttpRequest 类和 HttpResponse<T> 接口的常用方法分别如表 14-6～表 14-8 所示。

表 14-6　HttpClient 类的常用方法

方　　法	功　　能
static HttpClient newHttpClient( )	创建一个 HttpClient 对象，这个对象实现了 HTTP 客户端的基本功能
static HttpClient.Builder newBuilder( )	创建一个 HttpClient.Builder 对象，通过这个构建器，可以构建具有指定属性的 HttpClient 对象，以便后续使用这个对象进行 HTTP 通信。构建器模式是一种新型的对象构建编程模式，通过构建器可以更为灵活地设置所要构建对象的属性
abstract <T> HttpResponse<T> send (HttpRequest request, HttpResponse.BodyHandler<T> responseBodyHandler)	以同步方式发送 request 到指定的 HTTP 服务器，经由 responseBodyHandler 对服务器返回的结果进行处理后，作为 HttpResponse 对象返回。为了处理 HTTP 服务器返回的应答数据，HttpResponse 类提供一系列用于对应答进行处理的实现类
abstract HttpClient.Version version( )	获取 HTTP 的版本。目前有两个 HTTP 版本：HTTP 1.1 版和 HTTP 2.0 版

表 14-7　HttpRequest 类的常用方法

方　　法	功　　能
static HttpRequest.Builder newBuilder( )	返回一个 HttpRequest 对象 HttpRequest.Builder 构建器，通过这个构建器，可以构建具有指定特征的 HttpRequest 对象
static HttpRequest.Builder newBuilder (URI uri)	返回一个 HttpRequest 对象 HttpRequest.Builder 构建器，通过这个构建器，可以构建具有指定特征的 HttpRequest 对象。通过这个调用，设置 HttpRequest 对象的目的服务器地址为 uri 参数所指定的位置

表 14-8　HttpResponse<T> 接口的常用方法

方　　法	功　　能
T body( )	获取从服务器返回的应答数据的应答体
HttpHeaders headers( )	获取应答头相关信息
URI uri( )	返回给出响应结果的服务器的 URI 位置

为了灵活地构造 HttpClient 对象和 HttpRequest 对象，Java 的 HTTP 的客户端提供了 HttpClient.Builder 类和 HttpRequest.Builder 类两个重要的类。构建器编程模式是一种新型的对象构造模式，通过构建器可以更为灵活地设置所要构建对象的属性。关于 HttpClient.Builder 类和 HttpRequest.Builder 类，可以参见 JDK 的文档。

为了对从 HTTP 服务器返回的应答结果数据进行处理，在 HttpResponse<T> 接口中定义了一个重要的内部接口：HttpResponse.BodyHandler<T>，同时，对这个接口采用静态内部类做了基本实现，这个实现类就是 HttpResponse.BodyHandlers。

下面通过一个具体的例子说明如何使用 HttpClient 访问 Web 服务器的页面。通过这个例子，可以对 HttpClient.Builder 类、HttpRequest.Builder 类及 HttpResponse.BodyHandlers 有初步的了解。

14.4.2 使用 HttpClient 访问网络页面

这个例子使用 HttpClient 访问百度的首页。需要注意的是，这里只是将百度首页的 HTML 文档以文字形式显示出来，并没有做页面标签的解析。为此，在 ch14-01 工程下新建名为 com.ttt.http 的包，在这个包下新建名为 MyHttpClient 的类。这个类的代码如下（源代码为 14-06.java）：

```java
package com.ttt.http;
import java.io.IOException;
import java.net.URI;
import java.net.http.HttpClient;
import java.net.http.HttpRequest;
import java.net.http.HttpResponse;
import java.time.Duration;
public class MyHttpClient {
    public static void main(String[] args) {
        MyHttpClient myHttpClient = new MyHttpClient();
        myHttpClient.visitBaiduHomePage();
    }
    public void visitBaiduHomePage() {
        HttpClient client = HttpClient.newBuilder()
                .version(HttpClient.Version.HTTP_1_1)
                .connectTimeout(Duration.ofSeconds(20))
                .build();
        HttpRequest request = HttpRequest.newBuilder()
                .uri(URI.create("https://www.baidu.com/"))
                .build();
        HttpResponse<String> response = null;
        try {
            response = client.send(request, HttpResponse.BodyHandlers.ofString());
            System.out.println(response.statusCode());
            System.out.println(response.body());
        }
        catch (IOException | InterruptedException e) {
            throw new RuntimeException(e);
        }
    }
}
```

这个程序首先通过语句：

```
HttpClient client = HttpClient.newBuilder()
```

```
        .version(HttpClient.Version.HTTP_1_1)
        .connectTimeout(Duration.ofSeconds(20))
        .build();
```

创建了 HttpClient 对象。这个对象是通过 HttpClient.Builder 构建器创建的：设置使用的 HTTP 版本为 HTTP 1.1，设置访问超时时间为 20s，最后，使用 build( ) 方法构建 HttpClient 对象。之后使用语句：

```
HttpRequest request = HttpRequest.newBuilder()
        .uri(URI.create("https://www.baidu.com/"))
        .build();
```

构建了 HttpRequest 对象。通过构建器的 uri( ) 方法指定 Web 服务器的地址为百度的首页，然后通过 build( ) 方法构建了 HttpRequest 对象。之后再使用语句：

```
response = client.send(request, HttpResponse.BodyHandlers.ofString());
System.out.println(response.statusCode());
System.out.println(response.body());
```

将请求发送给服务器，通过 BodyHandlers 类的静态方法将服务器返回的结果转换为字符串，并将结果显示出来。运行这个程序，显示以下结果。

```
200
<!DOCTYPE html>
<!--STATUS OK--><html> <head><meta http-equiv= content-type content =
text/html;charset = utf-8 > <meta http-equiv=X-UA-Compatible
content=IE=Edge>
...// 省略了后续信息
```

从结果可以看出，程序获取了百度首页的 HTML 文档。

14.5　Java 网络编程应用实践

UDP、TCP 和 HTTP 都是非常常用的网络通信协议，它们有各自的使用场景：UDP 是无连接的点对点数据包协议，简单快捷，同时需要注意的是，默认情况下，一个 UDP 数据包能够发送数据的最大长度是 64KB；TCP 是面向连接的，采用 Java I/O 流对数据进行发送和接收，因此灵活，但是由于 TCP 通信需要客户端和服务器端进行三次握手才能建立连接，因此，效率略有损失；HTTP 被广泛应用于 Internet 网络应用，是 Web 应用的基础协议。

考虑到现代网络基础设施已经相当稳定可靠，因此，对于新设计的应用，可以首先考虑使用 UDP。

14.6 案例：聊天程序

微信、QQ 都是非常流行的实时通信聊天程序。本案例是设计一款简单的聊天程序，模拟微信群聊的效果。

14.6.1 案例任务

采用 UDP 实现一个简单的实时聊天程序，任何一个用户可以发送信息到参与聊天的所有人，同时，任何一个人可以接收到来自其他人发送的信息。

14.6.2 任务分析

由于需要进行多人同时聊天，也就是群聊，因此，需要设计一个聊天信息中心，它接收来自聊天用户的消息并转发给其他用户。因此，需要设计以下几个类。

（1）MessageCenter 类：聊天消息中心程序类，完成用户消息接收和转发功能。其中包含三个内部类：User 类，用于存储一个聊天用户的用户名和密码信息；Peer 类，用于存储聊天用户的 IP 地址和端口信息；MessageHandler 类，实现 Runnable 接口，用于处理从聊天客户发来的消息。

（2）Chatter 类：聊天客户端程序类，从键盘接收输入并发送给消息中心。

（3）Message 类：聊天客户端与消息中心进行消息通信的数据类。其中包含一个用于表示聊天信息类型的枚举类 MessageCode，消息类型包括登录消息、退出登录消息、普通消息、成功消息、失败消息。

（4）Utils 类：用于将 Message 对象作为数据，进行网络发送和接收的工具类。

14.6.3 任务实施

在 ch14-01 的工程下新建名为 com.ttt.chat 的包，在这个包下新建以下几个类：Message 类、Utils 类、MessageCenter 类、Chatter 类等。Message 类的代码如下（源代码为 14-07.java）：

```java
package com.ttt.chat;
import java.io.Serializable;
public class Message implements Serializable {
    private final MessageCode type;
    private final String data;
    public Message(MessageCode type, String data) {
        this.type = type;
```

```
        this.data = data;
    }
    public MessageCode getType() {
        return type;
    }
    public String getData() {
        return data;
    }
    public enum MessageCode {
        Login, Logout, Message, Success, Failed;
    }
}
```

Utils 类的代码如下（源代码为 14-08.java）：

```
package com.ttt.chat;
import java.io.*;
import java.net.DatagramPacket;
import java.net.DatagramSocket;
import java.net.InetAddress;
import java.net.InetSocketAddress;
import java.util.function.Consumer;
public class Utils {
    public static void sendMessage(DatagramSocket socket, Message m,
        String ip, int port) {
        ByteArrayOutputStream bos = new ByteArrayOutputStream();
        ObjectOutputStream oos;
        try {
            oos = new ObjectOutputStream(bos);
            oos.writeObject(m);
            oos.flush();
            byte[] data =bos.toByteArray();
            DatagramPacket packet =new DatagramPacket(data,0,data.length,
                                    new InetSocketAddress(ip,port));
            socket.send(packet);
        } catch (IOException e) {
            System.out.println("程序运行错误，错误信息如下：");
            e.printStackTrace();
        }
    }
    public static MessageDetail receiveMessage(DatagramSocket socket) {
        byte[] buf = new byte[1024*64];
        DatagramPacket packet= new DatagramPacket(buf,0,buf.length);
        try {
            socket.receive(packet);
```

```java
                String addr = packet.getAddress().getHostAddress();
                int port = packet.getPort();
                byte[] data = packet.getData();
                ObjectInputStream ois = new ObjectInputStream(new
                    ByteArrayInputStream(data));
                Object o = ois.readObject();
                return new MessageDetail((o instanceof Message m)?m:null,
                    addr, port);
            } catch (IOException | ClassNotFoundException e) {
                System.out.println("程序运行错误，错误信息如下:");
                e.printStackTrace();
                return null;
            }
        }
        public static class MessageDetail {
            public Message message;
            public String addr;
            public int port;
            public MessageDetail(Message message, String addr, int port) {
                this.message = message;
                this.addr = addr;
                this.port = port;
            }
        }
    }
```

Chatter 类的代码留作练习。MessageCenter 类的代码如下（源代码为 14-09.java）：

```java
package com.ttt.chat;
import java.io.IOException;
import java.net.*;
import java.util.ArrayList;
import java.util.LinkedList;
import java.util.List;
public class MessageCenter {
    private final List<User> users = new ArrayList<>();
    private final List<Peer> peers = new LinkedList<>();
    private DatagramSocket socket;
    public MessageCenter(int port) {
        try {
            socket = new DatagramSocket(port);
            users.add(new User("张三", "12345"));
            users.add(new User("李四", "12345"));
            users.add(new User("王二", "12345"));
            users.add(new User("马五", "12345"));
```

```java
        } catch (SocketException e) {
            System.out.println("程序运行错误, 错误信息如下: ");
            e.printStackTrace();
        }
    }
    public void go() {
        while(true) {
            Utils.MessageDetail md = Utils.receiveMessage(socket);
            new Thread(new MessageHandler(md)).start();
        }
    }
    private class MessageHandler implements Runnable {
        private final Utils.MessageDetail md;
        public MessageHandler(Utils.MessageDetail md) {
            this.md = md;
        }
        @Override
        public void run() {
            switch (md.message.getType()) {
                case Login -> handleLogin(md);
                case Logout -> handleLogout(md);
                case Message -> handleMessageTransfer(md);
                default -> System.out.println("收到了非法消息!");
            }
        }
    }
    private void handleLogin(Utils.MessageDetail md) {
        String[] s = md.message.getData().split(";");
        String name = s[0];
        String password = s[1];
        for(User u : users) {
            if ((u.name.equalsIgnoreCase(name) &&
                (u.password.equalsIgnoreCase(password))) {
                synchronized(peers) {
                    for(Peer p : peers) {
                        if ((p.ip.equalsIgnoreCase(md.addr)) &&
                            (p.port == md.port)) {
                            Message m = new Message(Message.MessageCode.
                            Failed, "登录失败");
                            Utils.sendMessage(socket, m, md.addr, md.port);
                            return;
                        }
                    }
                    Message m = new Message(Message.MessageCode.Success,
                    "登录成功");
```

```java
                    Utils.sendMessage(socket, m, md.addr, md.port);
                    peers.add(new Peer(md.addr, md.port));
                }
            }
        }
    }
    private void handleLogout(Utils.MessageDetail md) {
        String[] s = md.message.getData().split(";");
        String name = s[0];
        String password = s[1];
        for(User u : users) {
            if ((u.name.equalsIgnoreCase(name)) &&
                (u.password.equalsIgnoreCase(password))) {
                synchronized(peers) {
                    for(int i=0; i<peers.size(); i++) {
                        Peer p = peers.get(i);
                        if ((p.ip.equalsIgnoreCase(md.addr)) &&
                            (p.port == md.port)) {
                            peers.remove(i);
                            return;
                        }
                    }
                }
                Message m = new Message(Message.MessageCode.Success, "退
                出登录成功");
                Utils.sendMessage(socket, m, md.addr, md.port);
            }
        }
    }
    private void handleMessageTransfer(Utils.MessageDetail md) {
        for(Peer p : peers) {
            Utils.sendMessage(socket, md.message, p.ip, p.port);
        }
    }
    private static class User {
        String name;
        String password;
        public User(String name, String password) {
            this.name = name;
            this.password = password;
        }
    }
    private static class Peer {
        String ip;
        int port;
```

```
        public Peer(String ip, int port) {
            this.ip = ip;
            this.port = port;
        }
    }
    public static void main(String[] args) {
        MessageCenter mc = new MessageCenter(8888);
        mc.go();
    }
}
```

14.7 练习：完善聊天程序 Chatter 类的代码

第 14.6 节的聊天程序是很有代表性的通信程序。请仔细阅读该程序，完成 Chatter 类的设计实现。使程序运行后显示以下结果。

```
请输入登录名：张三
请输入登录密码：12345
请输入要发送的消息（按 q 键退出）:
Hello
请输入要发送的消息（按 q 键退出）:
收到消息：Hello
Ok
请输入要发送的消息（按 q 键退出）:
收到消息：Ok
q
```

第 15 章 使用 JDBC 访问数据库

如果程序需要存储大量的数据，同时需要对数据进行复杂的处理，那么数据库将是理想的用于存储和处理大规模数据的工具。为了对存储于数据库中的数据进行操作和管理，软件开发公司开发了数据库管理系统（database management system, DBMS）。目前常用的数据库管理系统有 MySQL、SQL Server、Oracle 等。Java 对访问和操作数据库中数据提供了很好的支持，这就是 Java 提供的 JDBC（Java data base connectivity）框架。本章对使用 Java 的 JDBC 框架访问和操作数据库中的数据进行介绍。

15.1 JDBC 概 述

JDBC 是 Java 提供的访问数据库的统一框架和标准接口。通过 JDBC，Java 程序可以访问和操作任何常见的数据库系统。Java 程序访问数据库的模型如图 15-1 所示。

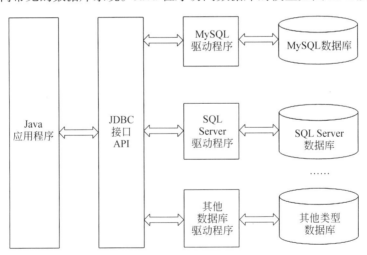

图 15-1 Java 程序访问数据库的模型

从图 15-1 可以看出，Java 应用程序通过 JDBC 接口 API（application program interface）和各个数据库厂商提供的 JDBC 驱动程序访问数据库。因此，在 Java 应用程序可以访问数据库之前，需要根据所使用的数据库管理系统不同，加载对应数据库管理系统厂商提供的 JDBC 驱动程序。

JDBC 接口有效屏蔽了数据库管理系统实现上的差异，从而隔离了应用程序与所使用的数据库管理系统之间的差异，提升了程序的适应性。例如，在开发 Java 应用程序的某个阶段可以使用 MySQL 数据库，在不需要或者只需对程序做较小修改的情况下即可访问 SQL Server 数据库，为程序开发带来了便利。

由于 JDBC 是一组标准的 Java 接口，因此，无论使用何种类型的数据库，访问和操作数据库中数据的过程都是一致的，总共包括以下几个步骤：第一步，加载数据库驱动程序；第二步，注册数据库驱动程序；第三步，获取与数据库的连接；第四步，访问和操作数据库中的数据；第五步，释放资源并关闭与数据库的连接。

15.2 加载数据库驱动程序

本书使用 MySQL 8.0 的社区版作为数据库管理系统，在使用 Java 程序访问 MySQL 数据库之前，需要安装 MySQL 8.0 数据库管理系统软件。由于安装 MySQL 8.0 数据库管理系统内容超出本书范围，在此不再赘述。作为本章的例子，利用数据库客户端软件创建名为 dbstudy 的数据库和名为 student 的数据表。创建数据库的语句如下：

```
CREATE DATABASE 'dbstudy'
```

创建数据表的语句如下：

```
CREATE TABLE 'student' (
    'id' bigint NOT NULL AUTO_INCREMENT,
    'name' varchar(100) NOT NULL,
    'addr' varchar(200) DEFAULT NULL,
    'photo' longblob,
    PRIMARY KEY ('id')
) ENGINE=InnoDB AUTO_INCREMENT=4 DEFAULT CHARSET=utf8;
```

然后向 student 表中插入少许数据。

为了加载数据库驱动，首先在 IDEA 中创建名为 ch15-01 的工程，然后，在 pom.xml 文件中增加以下内容。

```xml
<dependencies>
    <dependency>
        <groupId>com.mysql</groupId>
        <artifactId>mysql-connector-j</artifactId>
        <version>8.0.32</version>
    </dependency>
</dependencies>
```

新建的 ch15-01 工程如图 15-2 所示。

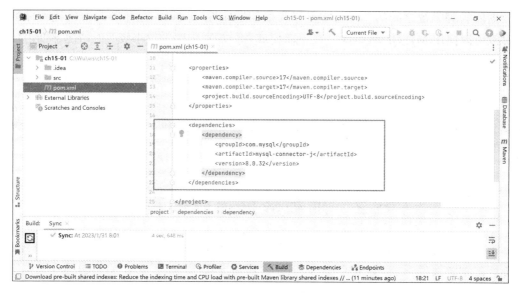

图 15-2 新建的 ch15-01 工程

在 pom.xml 文件中增加了指定内容后，单击图 15-2 中最下方的 Build 选项卡，然后单击左下方的 ⟲ 同步按钮，此时，IDEA 会将需要的 MySQL 8.0 的数据库驱动程序加载到工程中。

一旦在工程中加载了数据库驱动程序，即可使用 JDBC 提供的核心类和接口注册数据库驱动程序，并进行数据库数据的访问和操作。

关于 Maven 及其配置文件 pom.xml

15.3 JDBC 接口访问数据库的核心类和核心接口

使用 JDBC 接口访问数据库的核心类和核心接口都在 java.sql 包中，包括：DriverManager 类，用于注册和管理数据库驱动程序；Connection 接口，用于表示和管理与数据库的连接；Statement 接口和 PreparedStatement 接口，用于表示和抽象要执行的 SQL 语句；ResultSet 接口，用于表示和抽象数据库管理系统执行 SQL 语句后的结果。

15.3.1 DriverManager 类注册数据库驱动程序

DriverManager 类用于注册和管理数据库驱动程序。DriverManager 类的常用方法如表 15-1 所示。

表 15-1 DriverManager 类的常用方法

方　　法	功　　能
static void registerDriver(Driver driver)	注册指定的数据库驱动程序
static Connection getConnection(String url, String user, String password)	从 url 所指定的数据库，基于指定的 user 用户名和 password 密码获取一个与数据库的连接。通过这个连接，即可访问指定的数据库

方法	功能
static Enumeration<Driver> getDrivers( )	获取系统已经注册的数据库驱动程序
static void setLoginTimeout(int seconds)	设置与数据库建立连接时最长超时时间

除了可以使用 DriverManager 的静态方法 deregisterDriver( ) 注册数据库驱动程序外，还可以使用 Class.forName(String className) 注册数据库驱动程序。例如，MySQL 8.0 的驱动程序是 com.mysql.cj.jdbc.Driver，所以，可以使用以下语句注册 MySQL 8.0 数据库驱动程序。

```
DriverManager.registerDriver(new com.mysql.cj.jdbc.Driver());
```

或者

```
Class.forName("com.mysql.cj.jdbc.Driver");
```

15.3.2　Connection 接口建立与数据库的连接

Connection 接口抽象了与数据库的连接。只能通过 DriverManager 的静态方法 getConnection( ) 获取 Connection 接口的实现对象。Connection 接口的常用方法如表 15-2 所示。

表 15-2　Connection 接口的常用方法

方法	功能
Statement createStatement( )	创建一个基本的 Statement 对象，通过该 Statement 对象，可以执行 SQL 语句
PreparedStatement prepare Statement(String sql)	创建一个带参数的 PreparedStatement 对象。参数以 "?" 形式出现在 SQL 语句中，后续可以使用 set 方法设置参数的值
CallableStatement prepareCall(String sql)	创建一个可以执行数据库存储过程的 CallableStatement 对象
void close( )	关闭与数据库的连接
void setAutoCommit(boolean autoCommit)	设置是否指定提交数据库操作结果
void commit( )	提交本次数据库操作结果
void rollback( )	回滚本次数据库操作结果

例如，下列代码获取一个 Connection 连接对象，并在这个 Connection 对象上获取一个 Statement 对象和一个 PreparedStatement 对象，用于后续执行 SQL 语句。

```
Connection conn = DriverManager.getConnection(
        "jdbc:mysql://localhost:3306/dbstudy?" +
        "useUnicode=true&" +
        "characterEncoding=UTF-8&" +
        "serverTimezone=Asia/Shanghai&useSSL=TRUE",
```

```
            "root", "123456");
Statement statement = conn.createStatement();
PreparedStatement ps = conn.prepareStatement(
                    "select * from student where name like ?");
```

不同的数据库采用不同的 url 参数建立与数据库的连接。对于 MySQL 8.0，其 url 参数的一般格式如下：

```
jdbc:mysql// 服务器其地址：端口 / 数据库名？参数名 = 参数值 &...& 参数名 = 参数值
```

其中 "jdbc:mysql" 是 MySQL 固定的前缀。上面这个例子中，数据库服务器地址为 localhost，表示本机地址；端口为 3306；数据库名称为 dbstudy；对数据库中数据采用 Unicode 字符集；使用 UTF-8 进行编码；采用亚洲 / 上海时区；采用 SSL 连接形式。登录数据库系统的用户名为 root，密码是 123456。

在建立了数据库连接后，使用以下语句获取 Statement 对象和 PreparedStatement 对象。使用这些对象可以执行 SQL 语句。

```
Statement statement = conn.createStatement();
PreparedStatement ps = conn.prepareStatement(
                    "select * from student where name like ?");
```

第 15.3.3 和 15.3.4 小节会对 Statement 接口和 PreparedStatement 接口进行介绍。

15.3.3 Statement 接口执行 SQL 语句

Statement 接口是执行 SQL 语句的基本接口。可以从已经建立的 Connection 对象中使用语句 createStatement( ) 方法获得一个 Statement 对象。Statement 接口的常用方法如表 15-3 所示。

表 15-3 Statement 接口的常用方法

方法	功能
boolean execute(String sql)	执行 SQL 语句，并返回执行成功与否的结果
boolean execute(String sql, int autoGeneratedKeys)	执行 SQL 语句，并返回执行成功与否的结果。如果执行的语句是插入语句，则返回自增字段的值。后续可以通过 getGeneratedKeys( ) 方法获取自增字段的值
ResultSet executeQuery(String sql)	执行一条 SQL 查询语句，并返回查询结果
int executeUpdate(String sql)	执行一条 SQL 插入、修改、删除语句，返回被影响的数据条数
int executeUpdate(String sql, int autoGeneratedKeys)	执行一条 SQL 插入、修改、删除语句，返回被影响的数据条数。如果执行的语句是插入语句，对于数据表的自增字段，返回自增字段的值
ResultSet getGeneratedKeys( )	获取插入语句时自增字段的值
void close( )	关闭 Statement 对象并释放相关资源

例如，下面的语句将向 student 数据表插入一条记录，并获取自增字段 id 的值。

```
Statement statement = conn.createStatement();
statement.execute(
        "insert into student (name, addr) values('王二','天津市')",
        Statement.RETURN_GENERATED_KEYS);
ResultSet rs = statement.getGeneratedKeys();
if (rs.next())
    System.out.println(rs.getInt(1));
```

关于 ResultSet 接口的使用，将在第 15.3.5 小节中介绍。

15.3.4　PreparedStatement 接口执行参数化 SQL 语句

PreparedStatement 接口是执行参数化 SQL 语句的接口。可以从已经建立的 Connection 对象中使用语句 prepareStatement(String sql) 方法获得一个 PreparedStatement 对象。该接口的常用方法如表 15-4 所示。

表 15-4　PreparedStatement 接口的常用方法

方　　法	功　　能
void setInt(int parameterIndex, int x)	设置 SQL 语句中的 int 类型参数，参数索引从 1 开始
void setDouble(int parameterIndex, double x)	设置 SQL 语句中的 double 类型参数，参数索引从 1 开始
void setString(int parameterIndex, String x)	设置 SQL 语句中的 String 类型参数，参数索引从 1 开始
void setBlob(int parameterIndex, InputStream inputStream)	设置 SQL 语句中的 Blob 类型参数，参数索引从 1 开始
void setBytes(int parameterIndex, byte[] x)	设置 SQL 语句中的 Blob 或字节类型参数，参数索引从 1 开始
boolean execute( )	执行指定的 SQL 语句
ResultSet executeQuery( )	执行指定的 SQL 查询语句，并返回执行结果
int executeUpdate( )	执行 SQL 修改语句，包括 SQL 的 insert、update 和 delete 语句
void close( )	关闭 PreparedStatement 对象并释放相关资源

例如，下面的语句将从 student 数据表查询所有"张"姓的学生并显示查询结果。

```
PreparedStatement ps = conn.prepareStatement(
        "select * from student where name like ?");
ps.setString(1, "张%");
ResultSet rs = ps.executeQuery();
while(rs.next())
    System.out.println(rs.getString("name") + ", " +
            rs.getString("addr"));
```

15.3.5 ResultSet 接口处理查询结果

当执行的 SQL 语句是数据库查询语句，也就是类似"select..."这样的语句时，数据库会将查询结果返回给程序，此时，这些结果就是以 ResultSet 接口形式体现的。可以形象地将 ResultSet 理解为类似于表格的结果，其中包含一个指向当前查询结果集数据行位置的指针（也称为游标 Cursor），这个位置指针最初指向结果集第一条记录的前面。ResultSet 的结果如图 15-3 所示。

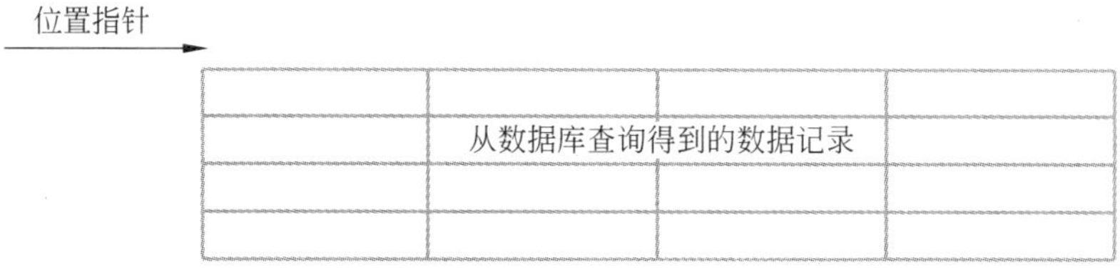

图 15-3 ResultSet 的结果

通过调用 ResultSet 的 next( ) 方法移动位置指针，当 next( ) 方法返回 true 时，表示位置指针指向了下一条记录；否则返回 fasle，表示到了结果集末尾，也就是说没有数据。ResultSet 接口的常用方法如表 15-5 所示。

表 15-5 ResultSet 接口的常用方法

方 法	功 能
boolean first( )	使位置指针指向结果集的第一条记录，成功则返回 true，否则返回 false
boolean last( )	使位置指针指向结果集的最后一条记录，成功则返回 true，否则返回 false
boolean next( )	使位置指针指向结果集的下一条记录，成功则返回 true，否则返回 false
Date getDate(String columnLabel)	从位置指针指向的记录中获取指定字段名称并且类型为 Date 类型的字段值
double getDouble(String columnLabel)	从位置指针指向的记录中获取指定字段名称并且类型为 double 类型的字段值
int getInt(String columnLabel)	从位置指针指向的记录中获取指定字段名称并且类型为 int 类型的字段值
long getLong(int columnIndex)	从位置指针指向的记录中获取指定字段编号的类型为 long 类型的字段值。结果集中的每个字段除了有和数据表一样的名称外，还有一个字段编号，从 1 开始
byte[] getBytes(String columnLabel)	从位置指针指向的记录中获取指定字段名称并且类型为 byte 类型的字段值
String getString(int columnIndex)	从位置指针指向的记录中获取指定字段名称并且类型为 String 类型的字段值
void close( )	关闭 ResultSet 对象并释放相关资源

第 15.4 节将通过一个完整案例介绍如何使用 JDBC 访问和操作数据库。

15.4 案例：实现对 book 表的增删改查

对数据库表的操作，总结起来无非就是向表中增加记录、删除表中的记录、修改表的记录和查询表的记录，简称增删改查。本节通过一个具体例子，介绍如何使用 JDBC 对数据库表进行增删改查操作。

15.4.1 案例任务

建立一个 MySQL 数据库，在其中创建一张 book 表，book 表包括以下字段：主键 id、书名 name、出版社 publisher、内容简介 memo、书的封面图片 cover。创建 book 表的 SQL 语句如下（源代码为 15-01.java）：

```sql
CREATE TABLE 'book' (
  'id' bigint NOT NULL AUTO_INCREMENT,
  'name' varchar(100) NOT NULL,
  'publisher' varchar(200) NOT NULL,
  'memo' longtext NOT NULL,
  'cover' longblob,
  PRIMARY KEY ('id')
) ENGINE=InnoDB AUTO_INCREMENT=6 DEFAULT CHARSET=utf8;
```

编写一个 Java 程序，实现对 book 表的增删改查，具体要求如下。
（1）可以从键盘输入要增加的记录数据。
（2）可以从键盘输入要删除记录的条件。
（3）可以修改指定的记录。
（4）可以从键盘输入查询条件，从而查询指定的记录。

15.4.2 任务分析

为了保存书的信息，可以定义一个简单的 POJO 类型的 Book 类。同时，为了实现对 book 表的增删改查，定义了 BookDAO 类，DAO（data access object）是一种与业务无关的数据操作类。为了实现业务功能，如从键盘输入数据等，再定义一个 BookService 类实现具体的业务功能。

15.4.3 任务实施

在 IDEA 的 ch15-01 工程下新建一个名为 com.ttt.bm 的包，在这个包下新建 Book 类、BooDAO 类、BookService 类。Book 类的代码如下（源代码为 15-02.java）：

```java
package com.ttt.bm;
public class Book {
    private long id;
    private String name;
    private String publisher;
    private String memo;
    private byte[] cover;
    public Book() {}
    public Book(long id, String name, String publisher, String memo,
    byte[] cover) {
        this.id = id;
        this.name = name;
        this.publisher = publisher;
        this.memo = memo;
        this.cover = cover;
    }
    public long getId() {
        return id;
    }
    public void setId(long id) {
        this.id = id;
    }
    public String getName() {
        return name;
    }
    public void setName(String name) {
        this.name = name;
    }
    public String getPublisher() {
        return publisher;
    }
    public void setPublisher(String publisher) {
        this.publisher = publisher;
    }
    public String getMemo() {
        return memo;
    }
    public void setMemo(String memo) {
        this.memo = memo;
    }
    public byte[] getCover() {
        return cover;
    }
    public void setCover(byte[] cover) {
        this.cover = cover;
```

```java
    }
    @Override
    public String toString() {
        return "Book{" +
                "id=" + id +
                ", name='" + name + '\'' +
                ", publisher='" + publisher + '\'' +
                ", memo='" + memo + '\'' +
                '}';
    }
}
```

BooDAO 类的代码如下（源代码为 15-03.java）：

```java
package com.ttt.bm;
import java.sql.*;
import java.util.ArrayList;
import java.util.List;
public class BookDAO {
    private Connection conn;
    public BookDAO() {
        try {
            DriverManager.registerDriver(new com.mysql.cj.jdbc.Driver());
            conn = DriverManager.getConnection(
                    "jdbc:mysql://localhost:3306/dbstudy?useUnicode=true&" +
                    "characterEncoding=UTF-8&" +
                    "serverTimezone=Asia/Shanghai&useSSL=TRUE",
                    "root", "123456");
        } catch (SQLException e) {
            System.out.println("程序运行错误，错误原因：");
            e.printStackTrace();
        }
    }
    public boolean insert(Book book) {
        try {
            PreparedStatement ps = conn.prepareStatement("insert into book " +
                    "(name, publisher, memo, cover)" +
                    "values (?, ?, ?, ?)", PreparedStatement.RETURN_
                    GENERATED_KEYS);
            ps.setString(1, book.getName());
            ps.setString(2, book.getPublisher());
            ps.setString(3, book.getMemo());
            ps.setBytes(4, book.getCover());
            int count = ps.executeUpdate();
            ResultSet rs = ps.getGeneratedKeys();
```

```java
            if (rs.next())
                book.setId(rs.getLong(1));
            rs.close();
            ps.close();
            return (count ==1);
        } catch (SQLException e) {
            System.out.println("程序运行错误,错误原因: ");
            e.printStackTrace();
        }
        return false;
    }
    public boolean delete(int id) {
        try {
            PreparedStatement ps = conn.prepareStatement("delete from
            book where id = ?");
            ps.setInt(1, id);
            int count = ps.executeUpdate();
            ps.close();
            return (count == 1);
        } catch (SQLException e) {
            System.out.println("程序运行错误,错误原因: ");
            e.printStackTrace();
        }
        return false;
    }
    public boolean update(Book book) {
        Book b = queryById(book.getId());
        if (b == null) return false;
        if (book.getName().isEmpty()) book.setName(b.getName());
        if (book.getPublisher().isEmpty()) book.setName(b.getPublisher());
        if (book.getMemo().isEmpty()) book.setName(b.getMemo());
        if (book.getCover() == null) book.setCover(b.getCover());
        try {
            PreparedStatement ps = conn.prepareStatement("update book
            set name=?," + "publisher=?, memo=?, cover=? where id=?");
            ps.setString(1, book.getName());
            ps.setString(2, book.getPublisher());
            ps.setString(3, book.getMemo());
            ps.setBytes(4, book.getCover());
            ps.setLong(5, book.getId());
            int count = ps.executeUpdate();
            System.out.println(count);
            ps.close();
            return true;
```

```java
        } catch (SQLException e) {
            System.out.println("程序运行错误,错误原因:");
            e.printStackTrace();
        }
        return false;
    }
    public Book queryById(long id) {
        try {
            Statement st = conn.createStatement();
            ResultSet rs = st.executeQuery("select * from book where id=" + id);
            if (rs.next()) {
                Book b = new Book();
                b.setId(rs.getLong("id"));
                b.setName(rs.getString("name"));
                b.setPublisher(rs.getString("publisher"));
                b.setMemo(rs.getString("memo"));
                b.setCover(rs.getBytes("cover"));
                return b;
            }
            rs.close();
            st.close();
            return null;
        } catch (SQLException e) {
            System.out.println("程序运行错误,错误原因:");
            e.printStackTrace();
        }
        return null;
    }
    public List<Book> queryByName(String name) {
        List<Book> list = new ArrayList<>();
        try {
            PreparedStatement ps = conn.prepareStatement("select * from book where name like ?");
            ps.setString(1, "%"+name+"%");
            ResultSet rs = ps.executeQuery();
            while (rs.next()) {
                Book b = new Book();
                b.setId(rs.getLong("id"));
                b.setName(rs.getString("name"));
                b.setPublisher(rs.getString("publisher"));
                b.setMemo(rs.getString("memo"));
                b.setCover(rs.getBytes("cover"));
                list.add(b);
```

```
            }
            rs.close();
            ps.close();
            return list;
        } catch (SQLException e) {
            System.out.println(" 程序运行错误, 错误原因: ");
            e.printStackTrace();
        }
        return list;
    }
}
```

BookService 类的代码如下（源代码为 15-04.java）：

```
package com.ttt.bm;
import java.io.ByteArrayOutputStream;
import java.io.FileInputStream;
import java.io.IOException;
import java.util.Scanner;
public class BookService {
    private final Scanner sc;
    private final BookDAO bookDAO;
    public BookService() {
        bookDAO = new BookDAO();
        sc = new Scanner(System.in);
    }
    public void go() {
        while(true) {
            String choice = menu();
            switch (choice) {
                case "1" -> add();
                case "2" -> delete();
                case "3" -> update();
                case "4" -> query();
                case "q" -> {return;}
                default -> System.out.println(" 非法选择! ");
            }
        }
    }
    private String menu() {
        System.out.println("1: 增加数据 ");
        System.out.println("2: 删除数据 ");
        System.out.println("3: 修改数据 ");
        System.out.println("4: 查询数据 ");
```

```java
            System.out.println("q: 退出程序");
            System.out.print(" 请输入: ");
            return sc.nextLine();
        }
        private void add() {
            System.out.print("\n 请输入书名: ");
            String name = sc.nextLine();
            System.out.print(" 请输入出版社: ");
            String publisher = sc.nextLine();
            System.out.print(" 请输入内容简介: ");
            String memo = sc.nextLine();
            System.out.print(" 请输入书的封面图片: ");
            String cover = sc.nextLine();
            Book book = new Book(-1, name, publisher, memo, null);
            byte[] buf = new byte[1024];
            try(FileInputStream fis = new FileInputStream(cover);
                ByteArrayOutputStream baos = new ByteArrayOutputStream();){
                int len;
                while((len = fis.read(buf))>0) {
                    baos.write(buf, 0, len);
                }
                book.setCover(baos.toByteArray());
            } catch (IOException e) {
                System.out.println(" 程序运行错误，错误原因: ");
                e.printStackTrace();
            }
            bookDAO.insert(book);
        }
        private void delete() {
            // 作为练习，请读者自行完成
        }
        private void update() {
            // 作为练习，请读者自行完成
        }
        private void query() {
            // 作为练习，请读者自行完成
        }
        public static void main(String[] args) {
            BookService bs = new BookService();
            bs.go();
        }
}
```

运行这个程序，程序将显示功能界面，然后按照提示输入数据，即可将数据插入 student 表中。

```
1：增加数据
2：删除数据
3：修改数据
4：查询数据
q：退出程序
请输入：1
请输入书名：Java 面向对象程序设计
请输入出版社：清华大学出版社
请输入内容简介：一本好书
请输入书的封面图片：C:/Wu/Temp/png-0010.png
```

15.5 JDBC 应用实践

客观上来说，使用 JDBC 操作数据库数据并不难，但是，要写出高质量的操作数据库数据的程序并不是件容易的事。在第 15.4 节案例程序中使用的分层结构是典型的数据库应用程序分层结构，这种程序结构可以有效隔离程序模块之间的耦合，降低程序设计编码和调试的难度，值得学习和借鉴。数据库应用程序的典型分层结构如图 15-4 所示。

图 15-4 中，Service 层处理与业务相关的功能，如在第 15.4 节的案例程序中，BookService 专门用来处理书籍数据的输入/输出；DAO 层负责完成与数据库的交互功能，并且是与业务无关的，如第 15.4 节案例程序的 Book DAO 负责处理对数据库的 book 表操作。Service 层与 DAO 层之间通过 POJO 对数据对象进行封装。

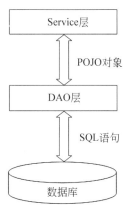

图 15-4 数据库应用程序的典型分层结构

15.6 练习：完成案例程序的删改查功能

完成第 15.4 节案例程序的 BookService 类对 book 表的删改查功能。

在第 15.4 节的案例程序中，留下部分功能并安排为练习的目的，就是希望读者能够通过通读案例代码理解应用程序的分层结构，并应用到项目实践中。

第 16 章　Java 图形用户界面

到目前为止，本书所有的例子程序都是基于文字界面的。其实，Java 提供了完整的用于开发图形界面的程序，也就是 GUI（graphic user interface）程序的 Swing 框架。本章对使用 Java 的 Swing 框架开发图形界面应用程序进行介绍。

16.1　Swing 概　述

图形用户界面是程序与用户交互的主要方式。Java 提供的 Swing 框架为开发图形界面程序提供了很好的支持。它包括完整的用于开发图形界面的组件和事件处理机制。Swing 框架包含的组件有：容器组件、功能组件、布局组件、菜单组件、工具条组件等；Swing 的事件处理机制使程序各个部分可以协调一致地工作。

考虑到 Swing 中组件类的方法比较多，同时，由于在编码时 IDEA 能给出详细的方法信息，因此，本章不再对 Swing 各个类的常用方法以列表形式做详细介绍。各个例子程序中用到的各个组件类的方法均是比较常用的方法，具有一定的代表性。对于其他组件类的其他方法，建议参考 Java JDK 17 的在线文档。

Java 图形界面：
Swing vs JavaFx

16.2　容　器　组　件

所谓容器组件，就是可以在组件中放置其他容器组件、其他功能组件的组件。例如，窗口就是一个容器组件，因为在窗口中可以放置包括按钮组件、文本输入框组件、菜单组件等在内的其他组件。Swing 中常用的容器组件包括 JFrame 和 JPanel。

16.2.1　JFrame 顶级窗口容器

JFrame 组件类似于 Windows 操作系统中的窗口，它是图形界面应用程序的顶级容器组件。JFrame 组件包括标题、窗口最大化按钮和最小化按钮、关闭窗口按钮等组件。应用程序通过直接创建 JFrame 类的对象或 JFrame 子类的对象构建应用程序的主界面。下面举一个简单例子说明如何使用 JFrame 类创建应用程序的主界面。为此，先在 IDEA 中新建一个名为 ch16-01 的 Java 工程，然后在 ch16-01 工程下新建名为 com.ttt.gui 的程序包。新建的 ch16-01 工程如图 16-1 所示。

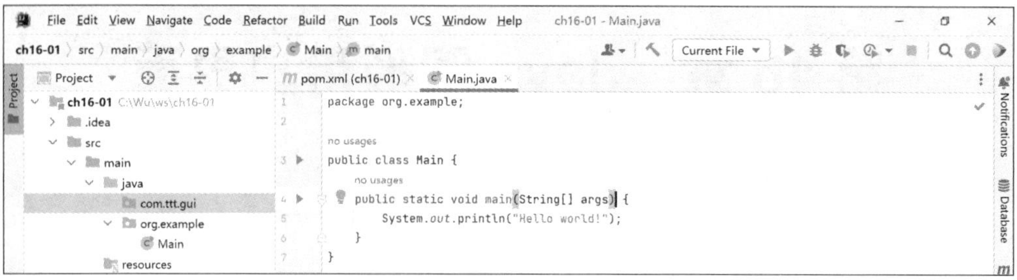

图 16-1 新建的 ch16-01 工程

在 com.ttt.gui 包下新建一个名为 MyJFrame 的类，这个类继承自 JFrame 类，因此是应用程序的主界面。MyJFrame 类的代码如下（源代码为 16-01.java）：

```java
package com.ttt.gui;
import javax.swing.*;
public class MyJFrame extends JFrame {
    public MyJFrame(String title) {
        this.setTitle(title);
        JButton button = new JButton("这是一个按钮");
        this.add(button);
        this.setLocation(300, 200);
        this.setSize(400,100);
        this.setDefaultCloseOperation(JFrame.EXIT_ON_CLOSE);
        this.setVisible(true);
    }
}
```

首先定义 MyJFrame 类是 JFrame 的子类。然后通过语句：

```java
this.setTitle(title);
JButton button = new JButton("这是一个按钮");
this.add(button);
```

设置窗口的标题，并将一个按钮作为组件添加在窗口中。通过语句：

```java
this.setLocation(300, 200);
this.setSize(400,100);
this.setDefaultCloseOperation(JFrame.EXIT_ON_CLOSE);
```

将窗口的显示位置定位在屏幕的 x 坐标为 300 像素、y 坐标为 200 像素的位置，并设置窗口的 400 像素宽、100 像素高，同时设置当单击窗口右上角的"关闭"按钮 × 时关闭窗口。最后，使用语句：

```
    this.setVisible(true);
```

将窗口显示在屏幕上。现在修改 com.example 包下的 Main 类为以下代码（源代码为 16-02.java）。

```
package org.example;
import com.ttt.gui.MyJFrame;
public class Main {
    public static void main(String[] args) {
        MyJFrame myJFrame = new MyJFrame("第一个图形界面程序");
    }
}
```

运行这个程序，显示如图 16-2 所示的界面。

如果在如图 16-2 所示的界面中单击"这是一个按钮"，按钮的颜色会发生变化，但是程序并不会执行任何其他操作，因为程序还未对按钮的单击动作做处理。在后续章节中会介绍如何监听按钮的行为并做适当响应。

图 16-2　MyJFrame 程序的运行结果界面

16.2.2　JPanel 面板容器

在复杂的图形用户界面中，经常需要按功能将界面划分为多个相对独立的显示区域，此时，可以使用 JPanel 对窗口进行划分。JPanel 也是容器组件，因此，在 JPanel 中可以放置其他组件，当然，也包括其他 JPanel 组件。例如，在下面这个例子中，使用两个 JPanel 将 JFrame 窗口划分为两个显示区域。为此，在 com.ttt.gui 包下新建一个名为 MyJFrameJPanel 的类，MyJFrameJPanel 类的代码如下（源代码为 16-03.java）：

```
package com.ttt.gui;
import javax.swing.*;
import java.awt.*;
public class MyJFrameJPanel extends JFrame {
    public MyJFrameJPanel() {
        this.setTitle("通过 JPanel 将窗口划分为两个子区域");
        this.setLocation(300, 200);
        this.setSize(400,150);
        JPanel main_panel = new JPanel();
        main_panel.setPreferredSize(new Dimension(400, 120));
        this.add(main_panel);
        JPanel up_panel = new JPanel();
        JButton button = new JButton("第一个按钮");
```

```
        button.setPreferredSize(new Dimension(390, 40));
        up_panel.add(button);
        main_panel.add(up_panel);
        JPanel down_panel = new JPanel();
        JLabel jLabel = new JLabel("Java图形界面设计", JLabel.CENTER);
        jLabel.setPreferredSize(new Dimension(390, 40));
        down_panel.add(jLabel);
        main_panel.add(down_panel);
        this.setDefaultCloseOperation(JFrame.EXIT_ON_CLOSE);
        this.setVisible(true);
    }
}
```

在 MyJFrameJPanel 类的代码中，通过语句：

```
JPanel main_panel = new JPanel();
main_panel.setPreferredSize(new Dimension(400, 120));
this.add(main_panel);
```

创建了 JPanel 面板容器，并设置面板的大小为 400 像素宽、120 像素高，并将面板添加到主窗口中。然后使用语句：

```
JPanel up_panel = new JPanel();
JButton button = new JButton("第一个按钮");
button.setPreferredSize(new Dimension(390, 40));
up_panel.add(button);
main_panel.add(up_panel);
```

创建了另一个 JPanel 面板对象，并将一个大小为 390 像素宽、40 像素高的按钮添加到这个面板中，最后，将这个新创建的面板添加到 main_panel 中。类似地，使用语句：

```
JPanel down_panel = new JPanel();
JLabel jLabel = new JLabel("Java图形界面设计", JLabel.CENTER);
jLabel.setPreferredSize(new Dimension(390, 40));
down_panel.add(jLabel);
main_panel.add(down_panel);
```

再创建一个面板，在其中添加了一个 JLabel 标签组件，在标签组件上居中显示一段文字，最后，将这个新创建的面板添加到 main_panel 中。现在，修改 com.example 包下的 Main 类为以下代码（源代码为 16-04.java）。

```
package org.example;
import com.ttt.gui.MyJFrameJPanel;
```

```
public class Main {
    public static void main(String[] args) {
        MyJFrameJPanel myJFrameJPanel = new MyJFrameJPanel();
    }
}
```

运行这个程序，显示如图 16-3 所示的界面。

如果将图 16-3 的界面拉宽，会发现界面发生了变化：两个面板并排显示了，如图 16-4 所示。

为什么会这样呢？因为这是"布局管理器"的作用：在没有指定容器使用的布局管理器时，容器默认使用称为"流式布局管理器"的机制在控制界面上组件的显示，而流式布局管理器就是以线性的方式从左到右从上到下一个挨着一个地放置组件。下面介绍 Swing 中常用的布局管理器。

图 16-3　JPanel 例子的运行结果

图 16-4　拉宽图 16-3 后的显示效果

16.3　布局管理器

布局管理器用于控制组件在容器中的显示位置。Swing 常用的布局管理器包括 FlowLayout 流式布局管理器、BorderLayout 边界布局管理器和 GridLayout 网格布局管理器。

16.3.1　FlowLayout 布局

流式布局管理器以线性方式从左到右从上到下一个挨一个地放置组件。其布局策略是：首先在容器中放置第一个组件，如果容器在水平方向还有显示空间，则将第二个组件放置在当前行；否则，将第二个组件放置在下一行，以此类推。下面举一个例子观察一下 FlowLayout 的布局效果。

在 com.ttt.gui 包下新建一个名为 MyFlowLayout 的类。该类是 JFrame 的子类，其中包含一个 JPanel 的面板，该面板使用 FlowLayout 对添加到其中的组件进行布局。这里，只简单添加了 6 个 JButton 按钮。MyFlowLayout 类的代码如下（源代码为 16-05.java）：

```
package com.ttt.gui;
import javax.swing.*;
import java.awt.*;
public class MyFlowLayout extends JFrame {
```

```java
public MyFlowLayout() {
    this.setTitle("6个JButton按钮的流式布局效果");
    this.setLocation(300, 200);
    this.setSize(230,120);
    JPanel panel = new JPanel();
    panel.setLayout(new FlowLayout());
    this.add(panel);
    for(int i=0; i<6; i++) {
        JButton jButton = new JButton("按钮" + (i+1));
        panel.add(jButton);
    }
    this.setDefaultCloseOperation(JFrame.EXIT_ON_CLOSE);
    this.setVisible(true);
}
```

在MyFlowLayout类的代码中使用语句：

```java
JPanel panel = new JPanel();
panel.setLayout(new FlowLayout());
this.add(panel);
```

创建了JPanel面板对象，并且使用FlowLayout对这个JPanel对象进行布局，然后，将这个JPanel添加到主窗口中。再使用语句：

```java
for(int i=0; i<6; i++) {
    JButton jButton = new JButton("按钮" + (i+1));
    panel.add(jButton);
}
```

向面板中添加6个JButton组件。现在，修改Main类为以下代码（源代码为16-06.java）。

```java
package org.example;
import com.ttt.gui.MyFlowLayout;
public class Main {
    public static void main(String[] args) {
        MyFlowLayout myFlowLayout = new MyFlowLayout();
    }
}
```

运行这个程序会显示如图16-5所示的界面。如果拉宽窗口，则显示如图16-6所示的界面。

图 16-5 流式布局初始界面

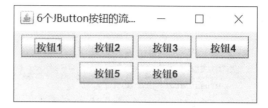

图 16-6 拉宽流式布局后的界面

16.3.2 GridLayout 布局

GridLayout 采用网格形式对容器中的组件进行布局。当向容器中添加组件时，按照先行后列的方式顺序放置到网格中。下面举一个例子观察一下 GridLayout 的布局效果。

在 com.ttt.gui 包下新建一个名为 MyGridLayout 的类。该类是 JFrame 的子类，其中包含一个 JPanel 的面板，该面板使用 GridLayout 对添加到其中的组件进行布局。这里，只简单添加了 6 个 JButton 按钮。GridLayout 类的代码如下（源代码为 16-07.java）：

```java
package com.ttt.gui;
import javax.swing.*;
import java.awt.*;
public class MyGridLayout extends JFrame {
    public MyGridLayout() {
        this.setTitle("6个JButton按钮的网格布局效果");
        this.setLocation(300, 200);
        this.setSize(230,120);
        JPanel panel = new JPanel();
        panel.setLayout(new GridLayout(2, 3, 4, 4));
        this.add(panel);
        for(int i=0; i<6; i++) {
            JButton jButton = new JButton("按钮" + (i+1));
            panel.add(jButton);
        }
        this.setDefaultCloseOperation(JFrame.EXIT_ON_CLOSE);
        this.setVisible(true);
    }
}
```

在 MyGridLayout 类的代码中，使用语句：

```java
JPanel panel = new JPanel();
panel.setLayout(new GridLayout(2, 3, 4, 4));
this.add(panel);
```

创建了 JPanel 面板对象，并且使用 GridLayout 对这个 JPanel 对象进行布局，然后，将这个 JPanel 添加到主窗口中。然后使用语句：

```
for(int i=0; i<6; i++) {
    JButton jButton = new JButton("按钮" + (i+1));
    panel.add(jButton);
}
```

向面板中添加 6 个 JButton 组件。现在，修改 Main 类为以下代码（源代码为 16-08.java）。

```
package org.example;
import com.ttt.gui.MyGridLayout;
public class Main {
    public static void main(String[] args) {
        MyGridLayout myGridLayout = new MyGridLayout();
    }
}
```

运行这个程序，显示如图 16-7 所示的界面。如果拉宽窗口，则显示如图 16-8 所示的界面。

图 16-7　GridLayout 初始布局界面

图 16-8　拉宽后的 GridLayout 布局界面

从运行结果可以看出，拉宽界面后，容器中的组件自动占满容器空间。拉高界面会显示类似的效果。

16.3.3　BorderLayout 布局

BorderLayout 布局，也称为边界布局，它将容器划分为东南西北中五个区域，如图 16-9 所示。在将组件添加到容器中时，需要指定组件将占用东南西北中的哪个区域。

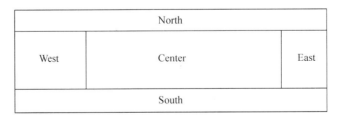

图 16-9　BorderLayout 布局

BorderLayout 布局经常被用来构建应用程序的主界面：在 North 区域显示程序名称；

在 West 区域显示功能列表；在 South 区域显示程序版权；在 East 区域显示功能说明；Center 区域则是工作区域。下面通过一个例子说明 BorderLayout 的使用。

在 com.ttt.gui 包下新建名为 MyBorderLayout 的类，该类实现一个典型的应用程序工作界面。MyBorderLayout 类的代码如下（源代码为 16-09.java）：

```java
package com.ttt.gui;
import javax.swing.*;
import java.awt.*;
public class MyBorderLayout extends JFrame {
    public MyBorderLayout() {
        this.setTitle("边界布局");
        this.setLocation(300, 200);
        this.setSize(600,300);
        JPanel panel = new JPanel();
        panel.setLayout(new BorderLayout());
        this.add(panel);
        JLabel label_north = new JLabel("我的图像处理系统", JLabel.CENTER);
        label_north.setPreferredSize(new Dimension(400, 40));
        label_north.setBackground(new Color(0xFF, 0xFF, 0xCC));
        label_north.setOpaque(true);
        panel.add(label_north, BorderLayout.NORTH);
        JLabel label_west = new JLabel("这里是菜单", JLabel.CENTER);
        label_west.setPreferredSize(new Dimension(80, 250));
        label_west.setBackground(new Color(0xCC, 0xFF, 0xFF));
        label_west.setOpaque(true);
        panel.add(label_west, BorderLayout.WEST);
        JLabel label_south = new JLabel("版权所有，盗版必究", JLabel.CENTER);
        label_south.setPreferredSize(new Dimension(400, 30));
        label_south.setBackground(new Color(0xFF, 0xCC, 0xCC));
        label_south.setOpaque(true);
        panel.add(label_south, BorderLayout.SOUTH);
        JLabel label_east = new JLabel("功能说明", JLabel.CENTER);
        label_east.setPreferredSize(new Dimension(60, 250));
        label_east.setBackground(new Color(0xCC, 0xFF, 0xFF));
        label_east.setOpaque(true);
        panel.add(label_east, BorderLayout.EAST);
        JLabel label_center = new JLabel("这里是工作区域", JLabel.CENTER);
        label_center.setBackground(new Color(0xFF, 0xFF, 0xFF));
        label_center.setOpaque(true);
        panel.add(label_center, BorderLayout.CENTER);
        this.setDefaultCloseOperation(JFrame.EXIT_ON_CLOSE);
        this.setVisible(true);
    }
}
```

在 MyBorderLayout 类的代码中，使用语句：

```
JPanel panel = new JPanel();
panel.setLayout(new BorderLayout());
this.add(panel);
```

创建面板，设置这个面板使用 BorderLayout 布局，并添加面板到主窗口中。然后使用语句：

```
JLabel label_north = new JLabel(" 我的图像处理系统 ", JLabel.CENTER);
label_north.setPreferredSize(new Dimension(400, 40));
label_north.setBackground(new Color(0xFF, 0xFF, 0xCC));
label_north.setOpaque(true);
panel.add(label_north, BorderLayout.NORTH);
```

创建一个 JLabel 标签组件，设置其大小为 400 像素宽、40 像素高，并设置这个标签的背景颜色为指定的颜色值。最后将这个组件添加到面板的 North 区域。其他几个区域的代码是类似的，不再赘述。最后，修改 Main 类为以下代码（源代码为 16-10.java）。

```
package org.example;
import com.ttt.gui.MyBorderLayout;
public class Main {
    public static void main(String[] args) {
        MyBorderLayout myBorderLayout = new MyBorderLayout();
    }
}
```

运行这个程序，会显示如图 16-10 所示的效果。

图 16-10 BorderLayout 布局效果

16.4 Swing 常用组件

Swing 提供了非常丰富的组件，这些组件中常用的组件包括 JButton、JTextField、JTextArea、JLabel、JList、JComboBox、JFileChooser 等。这些组件在 javax.swing 包下，它们都提供丰富的方法以控制组件的外观和行为，具体可参见 JDK 的文档。下面通过一个例子说明 Swing 常用组件的使用。

这个例子是以网格布局在一个窗口中显示常用组件的外观。为此，在 com.ttt.gui 包下新建一个名为 MyComponents 的类，其代码如下（源代码为 16-11.java）：

```java
package com.ttt.gui;
import javax.swing.*;
public class MyComponents extends JFrame {
    public MyComponents() {
        this.setTitle("Swing常用组件");
        this.setLocation(300, 200);
        this.setSize(500,300);
        JPanel panel = new JPanel();
        this.add(panel);
        panel.add(new JButton("按钮"));
        JLabel jLabel = new JLabel("", JLabel.CENTER);
        ImageIcon icon=new ImageIcon("C:/Wu/Temp/png-0010.png");
        jLabel.setIcon(icon);
        panel.add(jLabel);
        JTextArea jTextArea = new JTextArea(10, 20);
        jTextArea.setLineWrap(true);
        jTextArea.setText("可以输入一些文字");
        panel.add(jTextArea);
        String[] data = {"Java面向对象程序设计", "数据结构",
                         "算法分析与设计", "操作系统"};
        JList<String> myList = new JList<String>(data);
        panel.add(myList);
        JComboBox<String>comboBox=new JComboBox<>();
        comboBox.addItem("爱国");
        comboBox.addItem("敬业");
        comboBox.addItem("诚信");
        comboBox.addItem("友善");
        panel.add(comboBox);
        this.setDefaultCloseOperation(JFrame.EXIT_ON_CLOSE);
        this.setVisible(true);
    }
}
```

这个程序添加了几个常用组件到面板中。现在修改 Main 类为以下代码（源代码为 16-12.java）。

```java
package org.example;
import com.ttt.gui.MyComponents;
public class Main {
    public static void main(String[] args) {
```

```
            MyComponents myComponents = new MyComponents();
        }
    }
```

运行这个程序,显示如图 16-11 所示的结果。

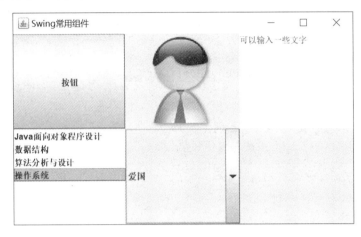

图 16-11 常用组件外观

因为没有对组件的显示外观做调整,同时未使用布局管理器对组件的放置进行处理,因此,如图 16-11 所示的界面还不美观。界面设计是一项非常耗时耗力的工作,需要耐心、审美和技术。

16.5 Swing 事件处理

Swing 的组件可以对鼠标事件、键盘事件等做出响应。例如,如果单击了某个按钮,程序应该对这个事件做出适当的响应。Swing 的事件包括:单击事件、光标移进或移出某个组件事件、键盘输入事件和窗口关闭、最大化、最小化事件。

Swing 事件处理需要设计三个关键对象:其一,Event Source,也就是事件源,触发事件的源点,如被单击的按钮就是事件源;其二,Event,也就是事件,用户对组件的一次操作称为一个事件,如键盘操作对应的事件类是 KeyEvent;其三,Event Handler,也就是事件处理器,它接收事件并对其进行处理,事件处理器需要实现某个事件的监听器接口。事件源、事件、事件处理器三者的关系如图 16-12 所示。

在组件可以响应事件之前,需要在组件上通过 add×××Listener( ) 方法(其中的 ××× 代表不同类型的事件)注册事件处理器,然后,当用户或者应用程序自身触发事件时,事件源会产生相应的事件对象,并作为参数传递给事件处理器进行处理。

下面举一个例子说明如何对组件的事件进行处理。在 com.ttt.gui 包下新建名为 MyEventHandler 的类。这个程序包含一个按钮、一个标签、一个文本框。当单击按钮时,在标签中显示一张图片;当光标移入或移出按钮区域时,在文本框中显示响应移入、移出

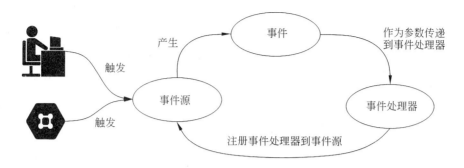

图 16-12 事件源、事件、事件处理器三者的关系

信息。MyEventHandler 类的代码如下（源代码为 16-13.java）：

```java
package com.ttt.gui;
import javax.swing.*;
import java.awt.*;
import java.awt.event.*;
public class MyEventHandler extends JFrame {
    private final JLabel image;
    private JTextArea ta;
    public MyEventHandler() {
        this.setTitle("Swing 事件处理 ");
        this.setLocation(300, 200);
        this.setSize(500,300);
        JPanel mPanel = new JPanel();
        mPanel.setLayout(new FlowLayout());
        this.add(mPanel);
        JButton btn = new JButton(" 单击按钮显示图片 ");
        btn.setPreferredSize(new Dimension(495, 50));
        btn.addActionListener(new MyActionListener());
        btn.addMouseListener(new MyMouseListener());
        mPanel.add(btn);
        image = new JLabel("", JLabel.CENTER);;
        image.setPreferredSize(new Dimension(200, 200));
        mPanel.add(image);
        ta = new JTextArea(" 鼠标动作: ", 12, 24);
        JScrollPane scroller = new JScrollPane(ta);
        scroller.setHorizontalScrollBarPolicy(JScrollPane.HORIZONTAL_SCROLLBAR_NEVER);
        scroller.setVerticalScrollBarPolicy(JScrollPane.VERTICAL_SCROLLBAR_ALWAYS);
        mPanel.add(scroller);
        this.setDefaultCloseOperation(JFrame.EXIT_ON_CLOSE);
        this.setVisible(true);
    }
}
```

```java
        private class MyActionListener implements ActionListener {
            @Override
            public void actionPerformed(ActionEvent e) {
                ImageIcon icon=new ImageIcon("C:/Wu/Temp/png-0010.png");
                image.setIcon(icon);
            }
        }
        private class MyMouseListener implements MouseListener {
            @Override
            public void mouseClicked(MouseEvent e) {
                String t = ta.getText() + "\n单击了按钮";
                ta.setText(t);
            }
            @Override
            public void mousePressed(MouseEvent e) {
                String t = ta.getText() + "\n按下了按钮";
                ta.setText(t);
            }
            @Override
            public void mouseReleased(MouseEvent e) {
                String t = ta.getText() + "\n释放了按钮";
                ta.setText(t);
            }
            @Override
            public void mouseEntered(MouseEvent e) {
                String t = ta.getText() + "\n光标进入了按钮区域";
                ta.setText(t);
            }
            @Override
            public void mouseExited(MouseEvent e) {
                String t = ta.getText() + "\n光标离开了按钮区域";
                ta.setText(t);
            }
        }
}
```

修改 com.example 包下的 Main 类为以下代码（源代码为 16-14.java）。

```java
package org.example;
import com.ttt.gui.MyEventHandler;
public class Main {
    public static void main(String[] args) {
        MyEventHandler myEventHandler = new MyEventHandler();
    }
}
```

运行这个程序，光标来回移进或移出按钮区域，或者单击按钮，将显示如图 16-13 所

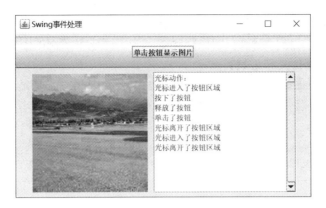

图 16-13　Swing 事件处理程序运行结果

示的界面。

16.6　Java 图形界面应用程序开发实践

一些软件企业使用 Java 的 Swing 框架开发图形界面程序。如果大家留意一下会发现，本书使用的 IDEA 集成开发环境就是使用 Java 及 Java Swing 开发的。使用 Swing 开发应用举例如下。

使用 Swing 开发应用运行举例

16.7　案例：图像混合器

图形混合器是一个有趣的应用程序：将两张图像混合起来，使混合以后的结果图像看起来具有原两张图形的视觉效果。

16.7.1　案例任务

采用 Swing 的 JFileChooser 组件选择两张图片，对图像进行混合，然后显示两张原图像和混合后的图像。

16.7.2　任务分析

Java 提供了较为完整的用于处理图像的工具包和相关类，包括 java.awt.image 包、

java.awt.Toolkit 工具类、javax.imageio.ImageIO 工具类等，通过这些类可以完成对图像的处理，如图像灰度化、图像缩放、图像降噪等。为了完成本案例的图像混合任务，需要用到 Java 的图像处理支持类和工具类。

首先考虑界面布局。由于要进行两张图像的混合，因此，需要允许用户选取两张用于混合的原始图像，并且需要将混合后的图像显示出来。因此，可以考虑如图 16-14 所示的界面布局。

选择原始图像	选择原始图像	图像混合
显示原始图像	显示原始图像	显示混合后图像

图 16-14　案例程序界面布局

因此，需要编写以下两个类。
（1）UIMixer 类：构建图形界面，并响应用户操作。
（2）ImageMixer 类：完成两个图像的混合。

16.7.3　任务实施

在 ch16-01 工程下新建名为 com.ttt.mixer 的包，在这个包下新建 UIMixer 类和 ImageMixer 类。UIMixer 类的代码如下（源代码为 16-15.java）：

```java
package com.ttt.mixer;
import javax.imageio.ImageIO;
import javax.swing.*;
import java.awt.*;
import java.awt.event.ActionEvent;
import java.awt.event.ActionListener;
import java.awt.image.BufferedImage;
import java.io.File;
import java.io.IOException;
public class UIMixer extends JFrame {
    private JButton btn1, btn2, btn3;
    private JLabel jLabel1, jLabel2, jLabel3;
    private BufferedImage image1, image2;
    private int width, height;
    public UIMixer() {
        this.setTitle("图像混合器");
        this.setLocation(300, 200);
        this.setSize(920,390);
        ui();
        this.setDefaultCloseOperation(JFrame.EXIT_ON_CLOSE);
```

```java
            this.setVisible(true);
            width = Math.min(Math.min(jLabel1.getWidth(),
                            jLabel2.getWidth()),jLabel3.getWidth());
            width = width - 30;
            height = Math.min(Math.min(jLabel1.getHeight(),
                            jLabel2.getHeight()),jLabel3.getHeight());
            height = height - 30;
        }
        private void ui() {
            JPanel mPanel = new JPanel();
            this.add(mPanel);
            JPanel buttonsPanel = new JPanel(new GridLayout(1, 3));
            buttonsPanel.setPreferredSize(new Dimension(900, 40));
            mPanel.add(buttonsPanel);
            ButtonActionListener bal = new ButtonActionListener();
            btn1 = new JButton(" 选择原始图像 ");
            buttonsPanel.add(btn1);
            btn1.addActionListener(bal);
            btn2 = new JButton(" 选择原始图像 ");
            buttonsPanel.add(btn2);
            btn2.addActionListener(bal);
            btn3 = new JButton(" 图像混合 ");
            buttonsPanel.add(btn3);
            btn3.addActionListener(bal);
            JPanel imagesPanel = new JPanel(new GridLayout(1, 3));
            imagesPanel.setPreferredSize(new Dimension(900, 290));
            mPanel.add(imagesPanel);
            jLabel1 = new JLabel("", JLabel.CENTER);
            imagesPanel.add(jLabel1);
            jLabel2 = new JLabel("", JLabel.CENTER);
            imagesPanel.add(jLabel2);
            jLabel3 = new JLabel("", JLabel.CENTER);
            imagesPanel.add(jLabel3);
        }
        private class ButtonActionListener implements ActionListener {
            @Override
            public void actionPerformed(ActionEvent e) {
                JButton b = (JButton)e.getSource();
                if (b == btn1) image1 = getImageName(jLabel1);
                else if (b == btn2) image2 = getImageName(jLabel2);
                else if (b == btn3) mix();
            }
        }
        private BufferedImage getImageName(JLabel lbl) {
            JFileChooser fc = new JFileChooser("C:/");
            int returnValue = fc.showOpenDialog(null);
            if (returnValue == JFileChooser.APPROVE_OPTION) {
                File iFile = fc.getSelectedFile();
```

```
            try {
                BufferedImage bi = ImageIO.read(iFile);
                Image im = bi.getScaledInstance(width, height, Image.
SCALE_DEFAULT);
                ImageIcon icon = new ImageIcon(im);
                lbl.setIcon(icon);
                return bi;
            } catch (IOException ex) {
                System.out.println("选择图像文件错误，原因如下：");
                ex.printStackTrace();
            }
        }
        return null;
    }
    private void mix() {
        if ((image1 == null) || (image2 == null)) return;
        ImageMixer imx = new ImageMixer(image1, image2, width, height);
        jLabel3.setIcon(new ImageIcon(imx.handle()));
    }
    public static void main(String[] args) {
        new UIMixer();
    }
}
```

ImageMixer 类的代码如下（源代码为 16-16.java）：

```
package com.ttt.mixer;
import java.awt.*;
import java.awt.image.BufferedImage;
import static java.awt.image.BufferedImage.TYPE_INT_BGR;
public class ImageMixer {
    private final BufferedImage bim1;
    private final BufferedImage bim2;
    private final int width;
    private final int height;
    public ImageMixer(BufferedImage bim1, BufferedImage bim2, int width,
int height) {
        this.bim1 = bim1;
        this.bim2 = bim2;
        this.width = width;
        this.height = height;
    }
    public BufferedImage handle() {
        BufferedImage bi1 = new BufferedImage(width, height, BufferedImage.
TYPE_INT_BGR);
        Image im1 = bim1.getScaledInstance(width, height, Image.SCALE_
DEFAULT);
        bi1.getGraphics().drawImage(im1, 0, 0, null);
```

```
BufferedImage bi2 = new BufferedImage(width, height, BufferedImage.
TYPE_INT_BGR);
Image im2 = bim2.getScaledInstance(width, height, Image.SCALE_
DEFAULT);
bi2.getGraphics().drawImage(im2, 0, 0, null);
BufferedImage bim3 = new BufferedImage(width,
                    height, BufferedImage.TYPE_INT_BGR);
for(int i=0; i<width; i++) {
    for(int j=0; j<height; j++) {
        int color1 = bi1.getRGB(i, j);
        int r1 = (color1 >> 16) & 0xff;
        int g1 = (color1 >> 8) & 0xff;
        int b1 = color1 & 0xff;
        int color2 = bi2.getRGB(i, j);
        int r2 = (color2 >> 16) & 0xff;
        int g2 = (color2 >> 8) & 0xff;
        int b2 = color2 & 0xff;
        int r3 = (r1 + r2)/2, g3 = (g1 + g2)/2, b3 = (b1 + b2)/2;
        r3 = r3<<16; g3 = g3<<8;
        bim3.setRGB(i, j, r3 + g3 + b3);
    }
}
return bim3;
}
}
```

运行这个程序，效果如图 16-15 所示。

图 16-15　图像混合器运行效果

16.8　练习：图形界面聊天程序

模拟微信群消息发送和接收功能，编写一个自制的图形界面聊天程序。要求：采用 Swing 图形界面方式实现该程序功能。

参 考 文 献

[1] 凯·S.霍斯特曼. Java 核心技术卷Ⅰ开发基础 [M]. 林琪，苏钰涵，译. 12 版. 北京：机械工业出版社，2022.

[2] 凯·S.霍斯特曼. Java 核心技术卷Ⅱ高级特性 [M]. 陈昊鹏，译. 12 版. 北京：机械工业出版社，2023.